They Served to Keep the Peace

By D. Ray Bowe

About the Author

D. Ray Bowe** grew up in rural Middle Tennessee early in the 20th century, at the end of World War II and during the height of the Cold War. There is no question that he is a country boy, but as a youngster watching a lot of military activity around his ridge-top home, his interest piqued in the Armed Services enough that he joined the National Guard. Later, he then enlisted in the United States Air Force and trained to be an aircraft engine specialist on the J-47 jet engine which powered the B-47 Stratojet, six-engine bomber. Yes, as a mechanic, he kept the bomber and some other fighter planes flying in tip-top shape, but Ray Bowe was not all warrior and soldier – he was also a man of God. His interest in God was so focused after his military career, that he went back to college and successfully completed two bachelor's degrees: one in Christian Ministry and another in Pastoral Studies. He went on to become an ordained reverend in the All-Saints Episcopal Church, and later an ordained priest.

> **All the names of the people in this book (either first, last, or both) have been changed to protect their identities, including that of the Author, himself. Some of the names of places or roads, etcetera have been changed as well.

Tribute to My Mom, Mary Evelyn Bowe

- Mother of Three Airmen That Put Her on Her Knees Daily -

Evelyn Bowe saw all three of her sons enlist in the military the same year - 1954. None of them even so much as gave her a goodbye hug or kiss when they left home for the service. As I now know, the Bosh and Bowe families were strange people. Us young men all probably left without thinking about when, and/or if, they would ever see their mother again. (As a family we weren't much for showing affection like hugging and kissing.) Probably the main reason for not saying so much as a word about even going to the recruiter, besides the trouble we were having with our dad, was that we just didn't feel like we would have an answer for all the questions she would ask. If we would have tried to say goodbye, she would have tried to talk us out of going but she also would have packed a lunch for the trip - that is the kind of mother she was. That is the kind of mother we had. With the spirit of the early pioneer housewife, Mary Evelyn raised the six of us; keeping us fed, clothed, schooled, protected, and policed. We children took all her skill, talent, and managing for granted. While we acknowledged her love for all of us as normal, she was however far and above the average housewife and mother. The many ways that she cared for and showed her love for us are interspersed throughout these pages.

As the oldest, I left home and school first. I thought my two brothers would help dad on the farm and continue as good players on the high school football team; however, they both followed suit within 6 months. Bud enlisted in the Air Force in the month of August 1954 and was home from Basic at Lackland AFB for Christmas. J.C. enlisted in December. During my adventures, brothers Bud and J.C. also had some of their own – although Bud received the more physical and dangerous assignment at first.

Table of Contents

My Birth and Early Years Previous

Though I wasn't born until 1934, my story was profoundly influenced by factors leading up to the end of the second decade of the 20th Century. In Russia, the autocratic Tsarist regime was badly beaten by Germany during the First World War, causing chaos which led to the Red Terror of the Bolshevik Revolution and the ultimate emergence of the new Union of Soviet Socialist Republic (USSR) in 1922. The United States of America entered WWI in 1917-1918, giving the near depleted and exhausted British and French armies much needed help - essentially cutting the throat of the German Army and bringing an end to the war on November 11, 1918, at the height of the Red Terror in Russia. Before long, the USA and USSR began competing with each other for world dominance - that competition became the Cold War. Each had an international mission to control the other.

Three things occurred all of which profoundly influenced the course of my life. First, the Cold War was a major influence on how I was shaped by the universal forces under God's sovereignty. Second was a spinoff of the Cold War but different; it was the sharpening of the ideological divide between communism and capitalism - two incompatible ways of life. Interestingly in the U.S., perhaps because of the war and severe economic depression, the Soviet alternative of communism attracted much support – not only here but around the world, becoming a focal point for its enemies and rivals. That ideological struggle shaped much of my life. Finally, the third and most important event (without which I wouldn't have existed), the birth of my father in 1913: Woodrow Wilson Bowe, known as 'Short Bowe, The Fox Hunter'.

My father was the oldest of seven siblings born to James Bird and Dora Carr Bowe. He never attended school, but he learned hard work on the farm training and using mules to pull the wagon, the turning plow, and other cultivating equipment. His siblings from oldest to youngest were: Bernice, Tommy (b. 1916) who was a WWII vet, Alton (b. 1924), James Junior, Dimple, and Willene. Both James Jr. and Alton were classified 4-F by the selective service, meaning that they could not be drafted into the armed forces. However, service to country ran in our blood and they both ended up in uniform, serving with the Civilian Conservation Corps.

My mother, Mary Evelyn, was born to James and Pearly Bosh in 1916 on Carr Ridge in Putnam County, Tennessee. Born in the depression years on a country farm 15 miles west of the county seat of Cookeville, TN, she grew up the hard way. Fortunately, she did get through the 6th grade at Hopewell Elementary School. Mom had three sisters (Minnie, Wincey, and Francis) and a brother (Thurston), all of which helped raise chickens and work the farm. She and my father continued to work and live with my mother's parents in the early years after their marriage in Boma, TN by the Justice of the Peace in 1932.

As a young man, my dad was very industrious. He was able to purchase a work truck and earned some money delivering for neighbors and acquaintances. Because the draft board had deferred him on account of the four children, he was also a logger hiring and training '4F-ers' and parolees while WWII was in progress. I was almost 7 years old when Pearl Harbor happened. It was a frightening and dangerous time!

1934

My name is Dorris Ray Bowe. My parents named me after Dorris Macon, son of David Macon (known to the music industry as Uncle Dave). Dorris and David were popular country music artists at the Grand Old Opry in Nashville, TN. I was born in a little cabin on a trail off Anderson Road and Carr Ridge near Indian Creek, in Putnam County, Tennessee on Christmas Day of 1934. We lived 14 miles west of the county seat in Cookeville, Tennessee. Like my father, I too am the oldest survivor of seven siblings. My older sister, Virginia Joyce, died at the age of 6 months. As the first male child born into the Bowe and Bosh families, I was a mean spoiled brat at 2 and 3 years of age no doubt as I received more attention than needed from my six aunts and four grandparents from both sides of the families! Eighteen months later, in 1936, my brother "Buddy" Hershel Bowe was born. Being close in age, we became unusually close-knit as brothers.

Early on, before our family consisted of more than four, we lived with either our mother's or father's parents during a few of those depression years from 1933 to 1937. Both the Bosh and Bowe families planted and cultivated large vegetable gardens to have raw and fresh vegetables to both eat and preserve in sealed cans over the summer in order to eat well in the winter months. They raised hogs for meat and had cows to milk. Chickens were also raised for meat and eggs and to barter for other goods with both the rolling store and some of the stores in town.

My parents and grandparents worked in the fields of corn, which was food for both animal and human consumption, and the tobacco fields (referred to as 'patches') that were grown as a money crop and then harvested and delivered to a warehouse or auction sale. As is common in blended families, our dad and his father-in-law often had quarrels and heated discussions as they worked together in the fields - Dad's smoking and Grandpa Jim's habit of stopping mid-field to comment on the outstanding features of his mules, are a couple of examples that come to mind. My parents rented small houses and moved around on some of the other ridges near my grandparents before purchasing the two-room house on Bowe Ridge from Ralph McBroom.

First Memories

My first memories as a very young child were centered around my grandparents. For instance, at Grandpa Bosh's place, when walking in the yard the chickens would peck at my feet and the goat would try to eat my coat sleeve. The cattle would wake me up bawling for food or some reason and the rooster did the crowing. Grandma would whistle most of the time and we could hear Grandpa hammering out in the blacksmith shop later in the day. Pa would not allow me near the shop, but my Aunt Francis would allow me to watch as she retrieved water from underneath the hillside with a line, bucket, and rope riding a cage down to a spring of fresh water. There was a crank with drum and handle to bring up good, fresh, cold water that we drank from the bucket out of a dipper (a cup with a long handle).

Over at Pa Bowe's place there was lots of fun with the old gander chasing me. It would pinch me which hurt and I remember trying to run to one of the grownups so they would chase it away. Just as over at my other grandpa's, the rooster would wake me up as he crowed. The pigs squealing was another sound I recall. My young uncles, Alton and James Jr., would have water fights at the freshwater well and sometimes spoiled vegetable fights, too. Pa and Ma kept them busy getting firewood or working in the tobacco patch. (All seasons - summer or winter - there was always work in the tobacco area.)

One sound I would hear in the mornings was Charley Edmonds' Model A backing out of the barn for the trip to town. The motor made a distinct sound like a hen collecting her baby chicks, "tacticity titter tic". On one special occasion my uncles' cousins, Clarence and Charles Bowe, came by and borrowed Pa's horse drawn buggy and took me for the ride of my life. One would pull like a horse or mule and the other pushed while making sure I stayed safe in the buggy. They finally became tired and lifted me out of the buggy and took it back to the big red barn where Grandpa's Model T was stored. Beside the big barn there was a cabin or small house where my parents and my brother and I slept; although we ate at the big house with the rest of the family - as did Aunt Mamie and Uncle Bernice along with Uncle Tommy. Bernice and Mamie had a small general store by the highway where they sold produce, tobacco, candy, etcetera that also housed their sleeping quarters.

My favorite spot was as high as I could climb up on the big gate by the barn. I would watch the birds and crows fly and sometimes fight with the hawks. The cloud formations were interesting to watch and I learned what a storm cloud appeared like - very dark. One day while up on that gate, an airplane came over flying low. What noise and the gate shook! I never tired of being there, but someone would always come and spoil my enjoyment and carry me to the house.

Mom the All-American Housewife

Being the wife of a poor subsistence farmer with several children to care for, my mother, Evelyn, stayed busy cooking, washing clothes, canning vegetables, milking the several cows, and feeding the dozen or more hogs, chickens and other cattle. They had to grow almost everything they ate. Chickens and eggs were bartered for sugar, flour, coffee, tea, salt, pepper and other spices; cloth for making shirts and dresses; and many other items. It was fun to watch and try to help her as a young child 3 and 4 years of age.

I have often wondered how she was trained or prepared to be a housewife under those strenuous conditions. For sure, she likely followed in the steps of her mother and mother-in-law; collecting their recipes and learning their "made from scratch" way of cooking and preparing meals and caring for her family. Mom loved to read, and she learned from experience how to care for babies and older children. She successfully brought us six children into this world.

Mom was a great person in her own ways, keeping something to eat on the table for six children and managing to keep us in clean clothes, such as they were. She had to battle dad on many occasions to keep groceries on the table and have shoes for us to wear to school. It seemed that my mother often had to be the referee, judge and jury to keep peace in the family. Not to sell my father short, because I did not walk in his shoes; he was doing his best during those tough years when there was no work for farmhands or loggers. Sometimes his truck would break as they all eventually do, but he would struggle and always come up with the money for the proper repair. He was a fair mechanic, situations and the times forced him to be. And, besides, he had to have a way to go find his hound dogs so he could continue hunting. I guess he needed some outlet from the family situation he had placed himself in.

Mary Evelyn Bowe

(L-R): Ray, Carol, Shirley and Buddy

Mom was a faithful Christian and loved to attend church meetings; she loved her family and some of us are pictured above in later years. In addition to teaching us boys practical things like how to milk the cows, she also taught us how to pray and give charitably from what God gave us.

This is the house that we lived in when my youngest sister, Shirley, was born in. I had electric put in after I joined the Air Force.

Mom and her sister

My parents never owned a cow or mule of their own until I was about 12 years of age

Mom's Family -

Mom was a good cook and raised chickens and a large vegetable garden.

Life on the Ridges and in the Hollows

Like exceedingly wrinkled skin on your fingers after a long, hot soak in the tub, the landforms throughout much of western Putnam County Tennessee are a maze of high convoluted ridges and narrow dark valleys called hollows. If you have ever seen the surface of an Osage orange fruit (locally known as hedge apple), then you might have some idea of what this area looks like from the air. Before there were roads carved through this jumble of ridges and hollows, the rivers and large streams were the main transportation routes for fur trappers and trading post traffic. They had to travel upstream from Nashville, turning into the Caney Fork, and then into streams through Buffalo Valley, branching off the river into Indian Creek and then into other streams originating from springs in the various hollows.

The roads that we walked to school and the store, and their names, tell revealing histories. Early on, the ridges were named after the family that first settled there - planting their homestead on available land. In general, wealthy people populated the fertile easy-to-farm flat and bottomland, leaving the rocky hardscrabble ridges for those who were less well-off. Old horse and buggy trails connecting ridgetop homesteads didn't have names at first, but people living there knew that it was the road to Tom Brown's place. Consequently, before long it became known as Brown Ridge, Brown Ridge Road, or Brown Holler Road, whatever the case may be.

One large stream came out of Riley Brown Hollow. Another major stream, from springs on Bowe Ridge Canyon, joined together at a confluence leaving enough flat land to construct two log houses and enough rich soil to garden and farm. There was enough fresh water and a large hunting area that a good hunter could bag plenty of small animals for food and fur. Certain hollows and ridge tops had enough flat land available to clear for tobacco, corn, and other crops, for large vegetable gardens, and even to sow in grass for grazing a few milk or beef cows. Between ridge top and hollow, the slopes were steep (too steep for growing) and best used for timber, hunting, and maybe grazing some stock.

There was, however, one house which had been built up in the northeastern portion of Brown Hollow canyon. As it was customary to name places after those who initially settled it, the Brown family was probably the first family to build there. My first recollection of the old house was the remnants of an old, rotting log foundation with some stones still standing at what used to be the four corners of the building. This dwelling, or at least what was left of it, was near two freshwater streams that came off the Robinson Ridge side. Riley Brown Hollow was just one of many valleys in this contorted area, but because of our acquaintance with a man named Ralph McBroom that lived there, we Bowe's referred to it as McBroom Hollow in the 1940's.

1941

Uncle Bernice Bowe had purchased two hillside tracts of land that joined the McBroom Hollow in various places near the creek. Our young family lived on Bowe Ridge, which is named after my great grandparents Bird and Tennessee Bowe who homesteaded many acres up there, near a big freshwater spring. We lived there when I was around ages 4-6, and I started to school during that time. As I grew up, vehicles were scarce, so we all walked. For a small child, it was an extremely long walk out to Old Walton Road. I recall having to trek the half mile or so out the ridge to Aunt Mamie's or Grandma Bowe's place which would have been only halfway to the school. They would walk with and carry me the rest of the way.

Dad and Fox Hunting

There was a great deal of fox hunting on the Bowe Ridge and that sport became very popular with my father, Wilson. He acquired dogs and loved to hunt the red fox that lived in the caves and crevasses of the bluffs, steep hillsides, and wooded hollows. I believe the foxes must have enjoyed the chase too because, even though they could hole up or go into a den for safety, they would instead attempt to outmaneuver the hounds, leading them over the most difficult terrain possible. Hunters would stay out on the top of those ridges all night; sometimes even as many as three nights, just to listen to those dogs' bark and run the grey and red foxes. The hunters enjoyed themselves immensely, erecting huge fires around which to eat and drink and tell stories. Those that didn't have to clock into a factory job would stay if those hounds were running.

Once, when I was still a boy, there was a sound down in the woods that concerned the adults. To my mom, it sounded like a woman screaming. The sound came from near our house on Bowe Ridge. Dad and a neighbor, Harry Fields, decided to check out the source of those sharp, loud screams. Dad gathered his dogs and encouraged me to come along. I was about 6 years old and terrified of the dark. My father had a habit of always spooking me and making me more afraid of the dark than I should have been.

Well, dad 'hissed' (sent) the dogs off in the direction of the screams. They were only gone for what seemed a short time and then returned. One of them was whimpering and carrying on somewhat, which caused the adults to make some frightening remarks about what the cause might be. In the meantime, the screams kept getting closer and closer. Dad had walked away as he tried to re-engage the foxhounds, who were never the fighting type, when the briars and weeds up on the bank began to move as Harry and I stood in the road. Becoming a bit nervous that it might be a big cat, the neighbor picked up a stone to hurl at this unseen creature. Even more frightened at this point, I tried to get into his pant leg and moved as close as possible. When he drew back to hurl the stone to keep the animal away, the back of his hand struck me in the side of the head and I went sprawling onto the road in the hard, cutting gravel. I was only knocked out for a short time, but the cuts and scratches stayed with me to prove that I had indeed been on a hunt. After that lick on the side of my head knocked me to the ground, I swore off fox hunting until my teenage years.

Over the years, I went on many hunts with my dad; sometimes riding mules and others walking. Some of the places Dad turned his dogs loose didn't have any roads, just a horse or mule trail. After the hunt, the foxhounds were trained to come back to the place where he had hissed them off. His older dogs would make it home within a few days, but the younger dogs or ones he had recently acquired had often not learned the direction home. (If lucky, they might remember the starting place where they were dropped off.) Depending on when it happened, dad would own 8-13 hounds at a time. He tried to keep close to half a dozen that were trained at all times, partially in hopes that the young dogs would learn to chase fox instead of deer or rabbit, but also so they would learn to return home with the older dogs.

I never understood why dad took so much interest in it. He and Uncle Bernice loved those hounds; they would go sit out in a field or near a woodland tree line and just listen to those dogs' bark while giving chase. Images of people getting dressed in "proper hunting attire" (such as riding boots, special trousers, and blouse), before mounting a horse and riding after their baying hounds as they chased after bushy-tailed prey, would be better suited to European-style fox hunting clubs. Tennessee hunters were just ordinary guys like my dad, who wore the same clothes every day: overalls, a white shirt, and a coat on top of that in cold weather.

Believe it or not, fox hunting was a popular sport in Tennessee – especially in the half dozen or so middle counties of the Upper Cumberland foothills. The Fox Hunters Association became large and well-known after I went into the military, so I wasn't around to know how good my father had become at it. Throughout the years, their week-long hunts and the field trials had earned my dad so many trophies that he had a shelf full of them by the time my sons were old enough to call my attention to those shiny mementos.

When we moved from Bowe to Thompson Ridge, I continued to attend Cedar Hill Elementary for a couple of years. The house that I first remember was a simple place - there was no running water. We didn't have an attached bathroom either, so we always went to the outhouse to do our business. In spite of the limitations, I remember happy years there. We had goats, and they sure did not like Momma. The first television that I remember was not at our house, but rather in the home of my aunt and uncle (after they were eventually able to afford one of these new-fangled things). Thompson Ridge was, of course, named after the Thompson family who lived nearby, and my brothers and I would go talk to them while they worked in the fields.

Another thing that stands out in my memory when I was new to the Thompson Ridge Community and walked to Cedar Hill school with the Mahan and Marrs children, was how they enjoyed picking on and bullying me. One day mom had observed the treatment I received and she punished me for not fighting back. My nature was such a moderate and peace-making one, but my parents were both fighters and they thought I ought to stand up for myself. I suppose if I would have given one of them a black eye the others would have stayed clear of me. Well, when we changed schools and I became a fighter, I received several spankings for it. My scuffles became so frequent that dad threated to give me another whipping after I had received one at school during the day.

My first car was a 1939 Chevy that I bought from my dad with money from the accident - Never could get the dog smell out of it but I'll also never forget my first date in it with Bev.

Dad, W.W. Bowe, was a fox hunter who took his hunting and dog training and trading seriously.

He received many trophies for his efforts.

The rest of the family kind of had a love-hate relationship with his dogs.

Dad was known as 'Short Bowe' the Fox Hunter

The hunters used an old school bus behind the barn for a hunting lodge when the weather was unfavorable.

Dad's cattle and dog truck

The thirteen dogs below seem ready for the hunt.

Mom wasn't happy about cooking for all of them.

1943

Eventually, my parents bought a two-room house back on Bowe Ridge and the McBroom's became our neighbors once again. I believe the home they purchased on three acres of land was built by Ralph and his father Oaks. Oaks and Jenny had six children. Ralph was the oldest, with Hallie, Huston, Lehman, Dimple and Dorothy coming after.

As we labored, often times we could hear and see Ralph McBroom working his mules a mile further along the Bowe Ridge, on the other side of the canyon on the hillside where the TVA electric power lines now run. Mr. McBroom was a poor, struggling sharecropper; wrestling out large tree stumps, big hard rocks, and constantly troubled by small animals eating his corn crops. At the time, the only transportation in and out of the hollows (besides on foot) was a horse or mule-drawn wagon or slide (similar to a wooden raft) in which I rode with Mr. McBroom and his brothers Lehman and Houston. You might say Ralph was a hold out in the hollow; he only hunted and trapped rabbits and squirrels for food and tried to raise tobacco for a money crop. Most of the rest of the families had gradually moved out to the ridges as the motor carriages began to travel the old Walton Road which is now Highway US70 North.

Walking the Ridges

There were two schools near us. For several years, all of us kids (except for the two older McBroom children), walked more than a mile west to attend Lafayette Elementary School. Most of the families and children attending that school were from Mathis Ridge. Figuring that beyond obeying the law that we had to attend somewhere, our parents were little concerned about our education, we quickly opted to ride the mile East to Cedar Hill when a Putnam County school bus began running a route. Oddly enough, it was us kids that made the decision to ride the bus and switch schools - not our parents. It was an easy decision for me, my brother, and our small sisters who were so tired of walking.

Miss Lena Plutnik was the principal at Lafayette Elementary. She had been dad's only schoolteacher, so he trusted her to not only teach his children correctly but to also make us behave. He did not know Dallas Leftwich, the principal at Cedar Hill, which made him quite concerned and uneasy with our choice of schools. I well remember my siblings and I being reprimanded and closely questioned about the change.

Cedar Hill turned out to be a good decision as this school was much more demanding and challenging. Principal Leftwich had high educational standards but thankfully he was both fair and tough in his dealings with us. Dallas could bring us rough farm boys to heel - it probably helped that he had a gruff voice and could come across as grouchy at times. Overall, he was kind and compassionate, a Christian that regularly attended church with his family. His son, Mickey, was one of my companions as we played backyard football together along with our brothers. Those were good days.

The area where we lived was known then as the Bethany Church Community. Living up on the ridge not too far from Bixby, on a still day with the breeze blowing the right direction, we could easily hear the daily train huffing and puffing up the steep Buffalo Valley grade to the top of the first plateau at Silver Point, then whistling as it came into the station at Bixby; stopping just long enough to drop off mail and pick up passengers for work in Cookeville.

Military convoys like this used to pass by us as we walked the mile home from school.

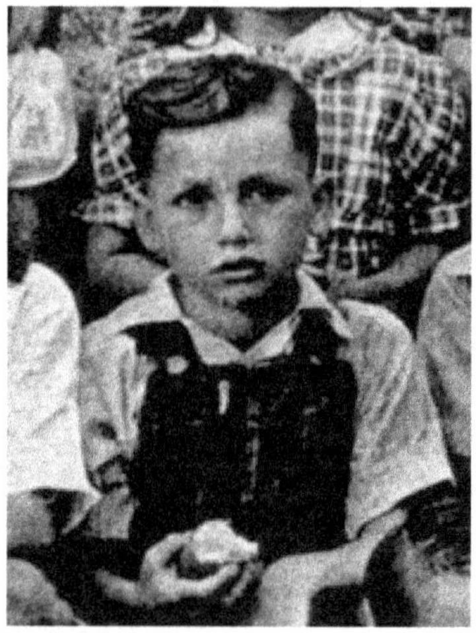

Young Dorris Bowe—7y.o.

8-year old Dorris Bowe is too small and young to really know what he wants, but as he watches the soldiers he is determined to be one of them some day.

Besides moving into our house, 1943 was eventful in that America entered the Second World War. Little did I know at the young age of nine how that and the following Cold War would shape my life in the military. I often wonder, *"Why did I go that way?"* What comes to mind is a cluster of mental images, perceptions, and sometimes even a story, which began to gradually, imperceptibly, shape my thinking and created the foundation for my future hopes and dreams to be in uniform, serving my country.

No doubt some of those images were real happenings in my early years as a grade-school student. I remember watching in wonder as military units would often pass us on the road as we walked to and from school. What impressed me were the orderly ranks of soldiers in uniforms so neatly dressed and fitted: boots shined, brass sparkling, trousers and jackets creased perfectly. They exuded strength in those attractive uniforms! I know my eyes became big and my heartbeat faster as I took it all in and imagined myself looking like that. They appeared as giants to me, and I wanted to be one of them.

Then later, as the war became more serious, Grandpa Bosh listened to the news on the radio with focused, intense concentration. I dared not make any noise while he tuned in to H.B. Caltenborn and Edward R. Murrow as they described the action on the war front. When I did forget and would come running through the room, Pa would angrily yell*, "Settle Down Jack!"* (He listened via a battery powered radio which had to be hand-cranked in order to recharge it.)

Yet, another early memory in grade school was watching the older boys drawing sketches and pictures of what they imagined the war to be like. Perhaps they saw some pictures on the cover of *Look* and *Time* magazines, or other news features. Their drawings depicted ships shooting tracers at airplanes and aircraft dropping bombs on ships. This, of course, was war as we thought it to be in our young imaginations. Those crude sketches were enhanced by living with the sights and sounds of military trucks and jeeps pulling trailers with Howitzers and anti-aircraft guns up and down US Highway 70N as we walked to and from school every day.

The soldiers would often throw candy and chewing gum to us kids as they drove or rode by. That candy throwing had one very interesting conclusion for my aunt one day, as someone in the passing vehicles tossed her a candy bar that had a note with his address wrapped around the treat. Private David Cope threw that candy bar to her. She must have liked both the sweet and the thrower, because before long they became sweet on each other and began writing many letters back and forth. The letter writing worked its magic and they ended up marrying after the war. David was from Pennsylvania, so Aunt Willene joined him there. At some point in her life, she became involved in The Church of the Nazarene, a faith tradition which we thought to be strange. I remember some stories about her Christian missionary work for that church. In fact, she was the first one I can remember who told me about Jesus Christ. She was a teenager when she shared the Gospel with me that first time.

The train went through my hometown daily at that time. Whoever might happen to be listening each morning could hear the train climbing the Buffalo Valley incline to the top of the first plateau leading to the Upper Cumberland Mountains. This same train whistle could also be heard when it reached Bixby.

Military Close to Home

Soldiers did more than just travel up and down our main road. There was one company of anti-aircraft batteries that came to the property our parents were renting on Thompson Ridge. The gun crew dug an earthwork emplacement in the field between our house and the old barn. Some of the soldiers would come over in the evening and listen to the radio, often swapping stories about fox hunting with my dad.

Many years later we learned the reason for all of the training done here in Putnam County and surrounding areas. As it turned out, these soldiers were destined for deployment to France and Germany which have terrain much like that of Middle Tennessee and Southern Kentucky. Training in the hills and valleys common to our area of the country gave the soldiers much needed practice in the concealment, navigation, and various other military maneuvers they would need in the regions they were eventually heading to.

Second cousins, Charles, and Clarence, decided to return home to Bowe Ridge to see their parents, Mr. and Mrs. Hobson Bowe, so we got to see them too. They sure were handsome in their uniforms. Others in uniform who attracted a lot of attention were Aunt Lennie's brothers Gilbert and J.C. All these young men, friends, and relatives in uniform were very impressive to this young grade-school boy - not to mention all of the other military activity going on close to home. Most of these soldiers were on home furlough before going overseas to fight and be wounded.

Carving

By the time I had reached the fourth grade at Lafayette Elementary School, I had somehow acquired a pocketknife. To a young boy, having a knife is a symbol of power and status; however, it can also prove to be a cruel implement until proper use and handling are learned. Mastering its use was often a dangerous and bloody process for me, involving many slips and mistakes. (The fact that I probably received the knife from a box of Cracker Jacks is an indication of the times we lived in.)

Trial and error taught me that a sharp knife works better than a dull one. Trial and error also taught me that soft wood with straight grain carves better than hard wood with a complex one. I made lots of mistakes cutting the wrong way against the grain. Finally, I learned how to apply the knife at the right angle and with the right pressure while always protecting my fingers. After far too many throw away pieces of whittled wood, I began carving out a small convoy of jeeps, trucks, and tanks. I had developed some drawing or sketching skills and eventually decided to draw my own version of the warfront complete with 3-D models. While my brothers had other interests, I became obsessed with carving and perfecting my creations.

Cousin Tommy Dale liked my version of the army vehicles, most all of which were carved out of yellow poplar. He was so fascinated with part of the convoy that one day he hid some in his pocket and took the pieces home to play with them there. Years later Tommy confessed to his larceny and asked forgiveness.

Apparently, there were so many that I didn't miss them, or my memory of that event faded. We both had a good laugh.

As my skills improved, I branched out, carving more complex pieces. One of my schoolmates and I began a friendly carving competition since we were both carving and building model military airplanes. He seemed to have a bit more skill, but both of us constructed planes with moving and detachable parts. I believe his warplane was a Corsair and mine was a Hellcat. Though neither of us had ever laid eyes on the real thing, we had seen numerous pictures of these classic warplanes.

Loss

The military happenings here in the Upper Cumberland were not *all* excitement, glamour, fun, and glory, however. Though news was generally very slow in arriving to these backwoods rural communities in the 40's, it did eventually trickle in. Tragic news traveled fastest, reaching some families by telegram and informing them a son was missing or, worse, killed in action. The awful reality and painful truth of war reached all the way down to us in the grade school at Lafayette Elementary.

We learned that some of the men from our community had been killed in action. One family, with other children who were my classmates at the time, got word of the death of their son who had recently been a student there. Upon hearing the terrible news, the teachers broke down sobbing and, through their tears, they described how they had taught these boys only three or four years earlier. Naturally, the younger brothers and sisters of those killed in action were badly hurt by the news and needed to be taken home. (Some, because of extreme grief or shock, had to have assistance.)

Cole Store

Just a hop, skip, and a jump from the small, two-room Lafayette Elementary School was the Cole Store. Not only was it a general store but also a gathering place for seniors and young folks who just wanted to hang out and hear the secondhand news. There was most always a pile of cedar shavings on or near the porch from those who sat there whittling as they talked. To my young mind, this old-timey general store managed by the Cole family contained everything imaginable. Yes, candy of all colors and stripes, toys, household items, and of course food. The store was a lifeline for those living nearby.

The Command Car

I am not sure whether or not the war was yet over, but at one point the American Allies had captured one of the German, bullet-resistant command cars and brought it to the U.S. to put on display as a war trophy. This car was driven throughout the country so the people supporting the war effort could see one of the spoils of war firsthand. And what a grand car it was!

We heard that it was coming our way; arriving in Memphis, then Nashville, and then coming East on what was the main Trailways bus route between Nashville and Cookeville in the 1940's - US Highway 70N. This was the road that we walked every day to and from school, so imagine our excitement to hear the siren and see the lights of that shiny, black Mercedes! The German military command car wound its way around the curvy road just as our classes let out and we were starting to make our way home. We couldn't have asked for better timing. With eyes shining and mouths agape, we watched in wonder as this glorious, never before seen car slowly drove past. What an experience for us grade schoolers! As youngsters, we couldn't possibly comprehend the importance or significance of what we had witnessed that evening as we casually walked along the only route home.

We Bowe kids had two Grandpa Jim's, the first being Jim Bosh: our mother's dad. Other than harvesting time, he wasn't around much. A natural craftsman, he had a blacksmith shop and expertly served farmers and railroad men as they brought or bought metal needing to be heated, bent, formed, and tempered into usable and functional items. Once our family moved to Bowe Ridge, we only interacted with this Grandpa Jim on occasional weekends when dad would make the 40-minute drive to Carr Ridge.

Mom was always ready to go and see her parents; a desire I came to understand much later after I had an out of state job and longed to see my family. My other Grandpa Jim was my father's dad, James Byrd Bowe - known to the family as Papa. Although he lived much closer, for some reason he did not spend much time with us kids either. We would see him only when there was a job, he needed help with; like when my father and I would go into the forest with him to cut down poplar trees to sell as paper pulpwood.

The Icy Creek

One cold and snowy day when I was about 9 years old, my dad covertly summoned me out to the porch as if briefing me for a top-secret mission. Turns out there had been a phone call from a friend who had spotted one of my father's hound dogs in the vicinity of a recent hunt. (It was not uncommon for one of them to get lost on occasion. The grey foxes would only run a short time before going into their dens, however the red fox seemed to enjoy the chase as it was usually able to "out-fox" the pack of hunting dogs when they lost track of the fresh scent.) Lesell Nichols had a farm down on Indian Creek Road near the Stanton Ridge. He had spotted our dog near his property and worried someone better come catch it before hunger could drive it to his chicken yard after eggs or end up killing one or more of his birds. Because Dad often seemed to care more for those dogs than he did for her or us children, my mother was pretty jealous of those hounds and would have been very angry to know he was about to send me out into bad weather on account of one of them. We all kind of despised those dogs for the same reasons.

The Nichols' place was about a mile-and-a-half from our home on the Bowe Ridge. Knowing my strength and determination despite my small skinny body, Dad had installed enough fear in me to trust that I would be obedient over any fears of danger or injury. My father instructed me to follow a road around the Bowe Ridge until I reached the end of the mail carrier route, take the wagon trail road up the steep hill, then continue up to and around the Marlow place to go through the cattle gate leading down the north side of the ridge. From there, I was to take the walking trail that led down the ridge point to the bridge that crosses the Indian Creek onto the road to the Nichols' farm. Well, not only was I strong and determined, but also stubborn like mother. (The Bosh's were all stubborn – especially grandpa and all my aunts.) Not realizing the reason for my father's detailed instructions, I decided to take a different trail that I knew would enable me to reach that walking bridge without having to climb that steep hill up to the Marlow place.

Unfortunately, whereas that route had been dry and manageable in the summer, the "little" creek was so swollen that I came to the point where a small boy such as myself couldn't walk or climb over. As I searched for a shallow enough place to cross, I slipped and fell into the icy water! Then, seeing as how my trousers were already wet, I decided I might as well wade on across. My young mind did not realize or know the full danger or power of swift-moving water, and this decision almost turned out to be disastrous as the current was so strong it nearly swept me off my feet. Only by the grace of God did I manage to reach the other bank of the creek and was able to climb out. Wet and cold, and knowing I couldn't go back, I continued across an open field; hardly able to walk the way my overalls were clinging to my legs. The larger creek still loomed between me and the main road leading to the Nichols' place. Now armed with a healthy fear of the water, I knew I must find a wide, shallow gravelly spot to cross in order to keep from being swept away to the river. By the time I finally located a suitable spot and successfully reached the other side, I was shivering uncontrollably and knew – even without a dog or news of one - I must find a place to dry out and head back home.

Thanks be to God, I spotted smoke coming out of a chimney from a house on the hillside a short distance down the road. Making my way up the cow path to what was normally an empty house, I was met by a strong black lady who basically carried me across the porch and through the door where the others inside were stripping and grading tobacco. Excited to see an almost drowned and half-frozen, wet young boy, she exclaimed, *"Goodness child, where in the world have you come from this time of day?" "You would have frozen to death in a short time!"* Grabbing every blanket and cover, she wrapped me up and began rubbing my arms and legs as the men began to remove my shoes, socks, overalls and Long John underwear - which were especially difficult to remove when wet. Fortunately, these folks were stripping tobacco in front of an open fireplace. They stretched bailing wire across the front of it and hung up my clothes to dry.

Feeling like they had been visited by an angel and glad they were able to help me, those people were overcome with joy! (Like the good Samaritan that helped the stranger who fell by the roadside leading to Jericho) Their kindness forever changed my perspective as I had not had much interaction with people of a different color prior to that occurance. I will always consider this an example of how God places people and things in the correct spots to preserve one of his chosen children. Once the clothes dried out and I appeared to be in traveling condition, I had given them enough information about where I lived on the Bowe Ridge that they were able to direct me to a safe route home. Arriving home as it began to get dark, the family was overjoyed to see me. After much hugging and many questions, Mom almost attacked Dad over the incident.

Between my 11th and 13th birthdays, I learned to work a team of mules, hitching them to a wagon or a turning plow. We also had to break or train the mules to be ridden. As they naturally did not want us perched on their backs, I took a few spills just like the cowboys in the rodeos. I also learned to operate other farm equipment. We cultivated Uncle Bernice's rough hillside farmland across the valley from the end point of Mathis Ridge and planted fields of corn for meal, animal food, and to trade. One of the fields was adjacent to or across from the property that I presently own.

Mom the Truck Driver

I never saw anyone struggle learning to drive the way my precious mother did. She had many things against her learning. First, she was in many ways physically uncoordinated. Second, her first driving lessons were in dad's old logging truck, and it didn't help that my father had very little patience – especially with mom. Yet another hinderance was that all the roads that she had to practice on were hilly, had lots of sharp curves both left and right, and most were slippery gravel. But that's not all! The truck that she had to learn to drive with had a stick shift on the floor and very loose, badly worn steering rods that led to seriously delayed oversteering. So, between her lack of fine-tuned coordination and the worn steering mechanisms, that truck was all over the road - eliciting a lot of laughter from us boys who had all learned to drive by the time our feet could reach the floor pedals. We tried hard to conceal our laughter in her presence, but when she couldn't see us, we laughed and laughed over her attempts to keep the truck on the road and the way she careened wildly from one side to the other, barely avoiding going into the ditch. While you couldn't say she ever learned to drive the old truck *well*, she did improve on the over-correcting somewhat. I was proud that Mom kept at it and knew it was because she had that one quality that brings most people to victory – winners never give up.

Sometime later, Dad acquired a 1939 Chevrolet which was much easier for her to drive; it had an automatic transmission with a shift lever up by the steering wheel. She became a good driver as she and Aunt Minnie Dyer would navigate those crooked roads to church at Chestnut Mound. Even now, I can shut my eyes and still see the image of that difficult learning experience that my mom went through.

The Poisonous Snake Bite

At 12 years old, I was large enough to accomplish a great deal of work on the farm under our dad's direction. School was out at Lafayette Elementary, and we were putting in our usual tobacco and corn crop for the summer. I had completed the 5th grade and was beginning to get the hang of school and the different subjects. I liked my teacher, Miss Lena Plutnik, who was also the principal there at Lafayette. That beautiful, two-room brick building was practically in the woods; built about a thousand yards from the main highway on a one lane road. When constructed, many trees had to be cleared away to make space for the building, playground, and the outhouses on each side. The men's toilets were located about 50 yards down a path between a tree and some shoulder-high bushes. The women were located on the other side, somewhat closer and in a more open area. This woodland area was, of course, infested with snakes of several different kinds and types. While most of them were harmless (such as the large, black Chicken Snake and what we called the 'Blue Racer'), there were also dangerous species like the extremely poisonous Copperhead. The teacher was constantly warning us, *"Watch for snakes!!"*

The school year had been over for about a month for us children and we were cultivating my uncle's steep hillside on the west hill at the very back end of the Bowe Ridge. Bernice had cattle and a thin strip of pastureland below the field where we were growing corn. The plants were up just the right height to weed for the first time through - which meant they would also be the perfect size and tenderness for cattle to eat should they get through the fence. While it was routine for them to go and check every other day in case any cattle had managed to get out, some thunderstorms had come through the evening before that had brought a hard, heavy rain, so he thought it best to drive around and check on the fence. A lot of rain had come down, so the creeks were running fast and out of their banks.

We noticed Bernice and Ed go by in the pickup truck, and it wasn't long before they returned with bad news. The rainfall had been so strong that the force of the water, rocks, and dirt going down the very steep hillside had created deep gullies in the field and washed out a section of the fence. Even some of the posts in the lower cornfield were down. Although the cattle had not yet found the opening, we knew it wouldn't be long before it happened. By the time we arrived, and even on his crutches, Bernice and Ed had worked their way down the thousand feet beside the creek. I had carried the axe down and dad (Short Bowe) had some wire and a couple of boards with him. Bob and Edward Calbert, who had come along to help, had already taken their shoes off to cross the creek. Not intending to pass up a chance to get my feet wet, I followed suit. My dad noticed and yelled out, *"Boys keep your shoes on! There is no time to play in the creek; we must get this work done!"* Some of the cows were already milling about toward the open space.

Finally, we all reached the area of downed fence to assess the damage. As there were at least three places that needed repair, Uncle Bernice started pointing out some broken branches and large limbs which had washed down that he thought we could use for props. I spotted a good-sized willow bush that looked large enough to work nicely. As I took the axe and struck at the base of the bush, *WHAM!* Something hit me on the left ankle. I looked down and yelled out, *"A snake bit me!!"* All hands-on deck immediately started my way. I was so frightened that I could only whimper when dad reached me. In a frightened voice he asked, *"Where did he get you?"* then noticed the snake trying to slither away. He grabbed the first solid thing he could find and struck and killed the Copperhead. Loudly, my dad notified the rest that it had bit me but didn't know where. *"On my left ankle!"* I cried. As I held my leg up, the blood began to flow and the area around the bite on my lower ankle started to change colors. As everyone looked on sorrowfully, Dad grabbed my arm and said, *"Let's go and get to help as soon as possible!"* All agreed, but just stood looking at each other and their bare feet! We had only gone a few steps before reaching the small stream of fresh water. As I looked at the blood streaming into the water, Dad rebuked me in an angry voice, *"I told you boys not to take your shoes off; now you can see why!"*

I was moving at a slow run, and dad was taking quick steps, as we continued to make our way up the trail toward the truck. Not in shape for this uphill sprint, he had to stop and catch his breath because the last 50 yards were very steep. At this point, dad had the presence of mind to pull his bandana from his pocket to wrap and make a tourniquet on my ankle. Interestingly, I remember seeing a jet fighter pass about 10,000 feet overhead. Although I had read about them in school, that was the first one I had seen or heard as they were still in the experimental stage.

Upon reaching the truck, we quickly got in and started the 15-mile trip to the Cookeville hospital - which at that time was not much more than an emergency aid station. However, on the way out, dad spotted Uncle James Jr. who was a practicing veterinarian. After my father informed him that I had been bitten by a Copperhead snake, Uncle Jr. looked at the ankle and promptly pulled out his pocketknife. After noting the horrified look on my face, however, he quickly folded it up and placed it back in his pocket and ordered, *"Get going as quickly as you can!"* Driving dangerously fast, we sped down the gravel road until we finally reached the highway, where he really put the "pedal to the metal" on the old pickup truck! I am not sure he even noticed the stop sign at Bixby crossroads.

Unfortunately, traveling that speed proved useless when we reached the hospital because the medics we informed did not approach the bite as very serious matter. They did ask several questions, including, *"How long ago, or what time did this happen?"* but then told us to wait in the waiting room for a doctor who would see me soon. It may have been 20-30 minutes, but it seemed like an hour. By the time he finally examined the bite and placed a wrap on the whole foot and ankle, the doctor told us that there was nothing he could do at this stage since the poison had already circulated through my bloodstream and reached all parts of my body. They took my vital signs and said, *"He seems okay."* The doctor looked at dad and said, *"Just watch him for any change."*

We left the hospital disappointed and beginning to wonder if this is how they deal with all patients. Our next thoughts were, perhaps next time we will go to the funeral home and make arrangements – at least then we might be shown some concern about how soon they could expect us. By the time we left the city limits, I began to hurt; it was not a throbbing pain like a broken bone or a toothache, but a tingling pain in many places: legs, shoulders, arms, and feet - many areas at once. I know dad was very angry at the hospital - and at me for taking my shoes off. The truth about it was that where the snake hit me would have been just above the top of my shoe. As I continued to squirm around in the truck seat, dad began to mumble under his breath. In an attempt at trying to subdue his real fears, he said, *"The next time, or if I had this to do over, I would purchase some corn liquor and have the person drink that. Perhaps that would make them so numb they would not be thinking about the results of cutting it open."*

There was a welcome party waiting when we arrived home. Everyone wanted to touch me as they were asking questions about what the doctor said and what were we supposed to do next. We were both very hungry and the mention of food brought out the beans and corn bread, little green onions, buttermilk potato salad, and iced tea. Mom, so concerned and happy to be able to do *something*, kept asking what else she could get us. Every family member wanted to know something or have input. Someone even tried to lighten the atmosphere by suggesting that the snake would have died from *my* poison venom if dad had not killed the thing!

Night came and, knowing we would be having visitors until late; I was given royal treatment – i.e., they allowed me to lay in the big bed in the living room. I don't remember who came except Aunt Mamie who unwrapped the leg and foot only to see nothing much more than a small puncture hole and a three-inch long scratch. It seems the snake must've struck me as I was in a turning motion so one fang entered and the other just dragged across. How that happened is history now.

In an unexpected visit, our neighbor Effy-Bell Wallace came by to inquire about the entire incident. I didn't know her; in fact, although we all knew who she was, none of us Bowe's had ever spoken with her before. She was different - her whole family was. All of us just watched and looked on with amazement as she proceeded to put her hands on and pray for me. I truly believe God may have used her touch and prayer to save my life! Although that leg began to swell up to twice the normal size and turned a brownish yellow over the next couple of days to the point where I couldn't walk by the third day, it did go back down, and I recovered. I was teased afterwards about Effy-Bell. She was a young married woman but kind of pretty in a certain way. Thank God for people who have a deep concern for human suffering and are brave enough to act on their faith.

Accepting Jesus

We were not yet a church family, Mom was saved first and soon afterwards there was a revival there at Bethany Church. Miss Lena, the schoolteacher who corrected my name, was the person who played and sang in church. She had taught us to pray and sing gospel songs at school, so she had a significant influence on my decision. As the youth leader, an Army Vet named Arnold Brown led the school age kids in song and worship as the Evangelist, Mrs. Booth, brought the message from the Bible. She read the story in the New Testament Book of Luke, chapter 19, about when Jesus entered and passed through Jericho and met a man named Zacchaeus who was chief among the publicans and rich. Paraphrasing the rest of the story: Zacchaeus was a very small man and could not see Jesus because of the taller people in front of him. So, he decided to climb a tree because he desperately wanted to see Jesus; as all men should! Being a small boy of eleven or twelve I could understand about tree climbing so the story had more meaning to me. This story changes many perceptions about Jesus and rich, sinful people as Jesus also sought Zach out and wanted to dine in his house. Jesus saw his heart, just as he sees ours, and wants to abide in our house regardless of our material wealth, social standing, political view or persuasions. The message is: Do not let him pass by; accept Him today without delay. Shortly thereafter, I publicly professed my faith and was baptized at the local Methodist Church.

Dissatisfaction

One day, while we were in the kitchen and Mom was cooking (she was a great cook), we were talking about the house we were living in. We had been there for perhaps eight years with no running water anywhere in the house - except perhaps at times when the roof leaked. I began complaining about the situation saying, *"When I leave home, I'm going to build a better home and make a good life for myself."* Mom began to cry and said, *"It will be a sad day, the day you leave home, and I do wish you well and want you to be able to have all those things; a shower stall or bathtub you can use each and every day."* Feeling ashamed of myself after her tearful response, I went to the cattle barn and started shoveling manure - because that was what I felt like after that selfish exchange. Although I wish I could say otherwise, there were more than a few similar exchanges like that before I left for the Air Force at age nineteen, but we at least had running water on the back porch by that time. (None of the houses we lived in had electricity or running water until I was in the 6th grade.)

Mom the Fire Fighter

One morning after Dad had gone to work logging, Mom and the rest of us were sitting around a warm fire in the wood burning stove. I was churning milk for butter by shaking a gallon jug of milk when Mom heard a roar and called my attention to the sound. Thinking it might be an airplane, I quickly put down the jug and ran outside to see what kind it might be. Looking up and around, I didn't see any planes, but what I saw instead was very frightening - THE HOUSE WAS ON FIRE!!

Flames and smoke were streaming from the wood-shingled roof! I quickly ran back into the house and yelled, *"Fire on the Roof!"* Mom quickly managed to get us children outside, then she went into action by climbing up into the attic and dislodging every board that was on fire. Those shingles were nailed together and also to the roof rafters, and I have no idea what she used besides her bare hands to dislodge them! By the grace of God, she was only slightly burned. Without a doubt, she saved that rental house on the Bowe Ridge and what meager furnishings we possessed at the time. Had this fire happened at night, we would have been history - gone and forgotten. God showed mercy and grace on this family because He had a plan for each and every one of us. He still does.

Running Away

Doesn't everyone run away from home at some time or another? I did. Mom and dad were struggling with us kids and each other. I was not really very focused in school and, although they needed to improve, there was no pressure or encouragement from any teacher to improve my grades or my prospects. Could that lack of focus have been caused by poor home conditions? After all, me and my two brothers still had to sleep in the same bed. Around age 14, and troubled because my parents were constantly quarreling with each other for hours after we went to bed, I became very unhappy and started causing problems. Our rooms were so close together, and the walls so thin, that we couldn't help listening to all that was said, shouted, cried, and bellowed. Perhaps I thought sooner or later I was going to become part of their constant fighting. Regardless, I had reached that rebellious stage where it seemed I just didn't have things I wanted, nor could I do things that I wanted to do.

I just couldn't help thinking that somehow I was being short changed. Why couldn't I be in another location, in another family? I often thought, "I don't want to be a Bowe anymore." Why couldn't I be doing chores other than what I was told to do? For example, the fencing around the lot where we kept the cow and mules was often down or nonexistent and animals would get out on the neighbor's side. Because this was usually discovered at night, us boys had to get out of bed, round up the animals and find the opening, and do some temporary patch work. The next morning, when it was light, we would be back at it strengthening the patch.

Usually there is some outside influence that contributes to the thoughts for runaways and misbehaving. Our neighbors the McBroom's were always more than willing to make a bad situation worse. Lehman McBroom, a playmate probably three years older and always full of mischief, suggested that we hitch a ride to Nashville with his sister, Halley, and venture over to the produce shelters and inquire about work. Another alternative was to go to Florida and pick fruit.

Undoubtedly, I was at that time in life when your body and mind go through a rapid growing stage, producing confused thinking. Though Mom couldn't talk to me about her own struggles, she was very concerned about me and turned out to be a good listener who was always there for me. Dad wasn't much help as it seemed he could care less about anything but his dogs. Although I was fourteen, I was only in the 7th grade due to the many seasons that my dad had kept me out of school (not just me but my brothers as well) when he needed help in the fields.

High School

Time and adventure changed my perspective as autos, girls, football, and studies took precedence as I entered high school. Life now became more complicated with the wants and desires associated with the beginning of manhood. The need for money and other provisions weighed heavily on top of my responsibilities. Childhood games with younger brothers now became uninteresting. As the oldest, I had privileges and responsibilities, one of which was caring for my younger siblings - especially my youngest sister Shirley who was a newborn when I was in my first year at the local high school.

Mom the Coat Maker/Seamstress

When I was a freshman in high school, autumn had passed, and a killing frost warned of a cold winter coming to the Upper Cumberland where we lived. As I had outgrown the winter coat that kept me warm the year before, I no longer had a decent one to wear to school so Mom and I came up with a plan to make me a navy blue ‘P Coat’ based off of a picture she had seen in a catalog. She located the material to modify or completely reconstruct what had perhaps been a lady’s overcoat. We imagined that material would do to make one of those coats that had a large high collar when turned up.

Mom was great with that sewing machine; she could take a dress pattern and make all the girls some nice ones. It was a great coat except there were enough mistakes that it was obviously not store bought and the guys at school asked me about it. Like a fool, I was embarrassed instead of bragging that it had been handmade for me by my mother. But homemade or not, it kept me warm like a blanket when we had to stand in the cold wind and wait for the school bus! I’ve often wondered how the Amish kids got around the teasing and unfair remarks I endured. I guess they attend all-Amish schools where almost every stitch of clothes the kids wear is homemade. Mom sure loved and tried to protect me from the elements. She tried hard to treat us all the same but, as I suppose happens in all families, I was her favorite and Bud was dad’s.

Edward Keaton came to live with Uncle Bernice and Mamie at age 9 and was a great help to them; rounding up and loading/ unloading young calves, etc. He attended Lafayette Elementary School with my brothers and I, but entered high school at age 12 because he was so smart.

Ed took possession of Bernice's truck and worked hard in the tobacco patch and vegetable garden. He graduated from high school at 16 and was rewarded with a Ford car.

Pee-Wee Keaton

None of my stories about high school are complete without identifying Pee-Wee because he was such an important part of our home, family, and school life in so many ways. Perhaps because he lived with Uncle Bernice and Aunt Mamie, I always thought of him as a cousin. Edward (Pee-Wee) Keaton came into our family when he was 9 years old. His mother was a single parent and he had a 12-year-old brother, Will Junior, with whom he argued constantly. Apparently, Mrs. Keaton had decided Eddy was more than she could handle or afford to support and had determined to place him in a home or reform school. As a result of her threat, Eddy started hitching rides with my uncle, Bernice Bowe, going to the livestock sale. Having contracted Polio as a child, my uncle walked with crutches. Being a "never give up" kind of guy, he had learned to live with and work despite his disabilities, raising tobacco and trading small calves. He decided this spunky third grader needed a home and that he and his wife could use the help around the farm. (It was a nice match for all three.) My Aunt Mamie had been barren up to that point, so they didn't have any children. She worked at the shirt factory and needed Eddy's help around the property.

Although in some ways he had to grow up sooner than other boys, up to this point in Ed's short life he was never given any true responsibilities. So, when he was sort of given charge of caring for the house and yard, he took this new responsibility very seriously and, sadly for us, maybe a little too much so where it concerned their pickup truck. My brothers (Bud and Calbert) and I were accustomed to playing inside it as small boys did in those days. Ed, however, became very possessive of the truck and even got mean if one of us so much as thought of sliding behind the steering wheel. At first, we as children resented the new addition to the Bowe family. After all, my brothers, my sister, and I had become accustomed to all the love and affection that Aunt Mamie and Uncle Bernice showered on us since they had given up trying to have their own children. Why would we want to share all that attention with an interloper?

Eddy was a smart kid but, being born and raised for his first eight years in the city of Nashville, had been allowed to smoke, drink, and use filthy language. All of that ended when he came into the Bowe home. He sure was a skinny kid, but even so he was strong for his size. (Like the rest of us, he needed to be fed better.) All the kids in our school and community were thin; we lived on beans and potatoes, corn bread and cow's milk. Pee-Wee attended Lafayette Elementary School with all the other Bowe and Gentry kids.

We didn't know what the teachers would think of Edward and his city-wise attitudes. Not only was Pee-Wee street-smart, but he was also a fighter. He'd had so many fights with his brother, Junior, that he knew how to throw a punch – plus a big mouth that boasted of just what he could do - to go with it. My brother Bud, who was the same size but a couple of years younger, disagreed with something Ed said one day and promptly received a sucker punch. When I objected and pointed out that we had been warned about fighting, Ed was so fast he gave me a couple of lefts before I could even put up a guard!

Well, the next day at school as he was bragging about beating up on the Bowe brothers, the McBroom boys (who thought that job was theirs) took offense and decided they wanted to see for themselves if this little guy could beat me up. Having seen and heard their plan, and knowing the McBroom's were rotten to the core, I hastily walked on ahead hoping to avoid a fight. Lo and behold here comes Ed up behind me, stepping on my heels and kicking my legs trying to trip me. Somehow I caught Ed off balance and, hitting him with a right cross, he went to the ground. Before he could recover, I put him down for the second time and had him crying by the third time. Our teacher, Miss Lena, heard about part of what had happened. Not bothering to learn the facts, she paddled all four of us - but I got the worst of it because I had fought with another kid in school before.

Like I mentioned previously, Pee-Wee was smart. The teachers, recognizing his talent and intelligence level, gave him some tests and he was able to skip through several grades. They gave him the 5th grade books and persuaded him to study so the next year Ed was able to come back as a 6th grader. He also did well working with Uncle Bernice, whom he actually listened to and respected, keeping up just like a man. Mrs. Lena administered the 6th grade test there in the two-room school, which he passed. The teachers continued to work with him the next year and, when they gave him the 7th grade test, he passed that as well. Ed was able to enter high school as a 12-year-old, and this was the reason for receiving the nickname "Pee-Wee". Though he was a small kid even for his age, he seemed even smaller when compared to the other high schoolers. I could see his classmates looking over their shoulders and sneering, "Who does that little kid belong to anyway? What is he doing here?"

Pee-Wee graduated with honors at 16 and was offered a scholarship but he turned it down because my uncle needed him. He had come to love Bernice as much as the rest of us. Although he was a Keaton, since he lived with my aunt and uncle, it felt like he represented our family. Boy were we proud that one 'Bowe' knew something; could do arithmetic, spell words and read well!

Thurston Neil

Thurston was one of my friends in high school and the first one of us to obtain a car. One night as my brothers and I were out riding around in his car, we were stopped by a Tennessee State Patrol officer by the name of Red Joly who knew the four of us from the community. Perhaps he had originally pulled us over for something unlawful like a taillight out, or maybe he even suspected where we had been, but intoxication was involved, and I slid any remaining beers under the driver's seat. After Patrolman Joly confiscated his driver's license, Neil looked over at me and said, *"Dog, you drive us home."* (Even though I didn't have my license yet, he knew I had enough driving skills to get us home that night.) A very nice, thoughtful man, the officer came and sought Thurston out the next day to return his license and made the comment, "Boys must experiment with the wild side of life."

Joining up - July 1952

World War II ended and the soldiers were returning home, but the Korean Conflict was heating up. The war over there was first called the French Indo-China War. While I was still in grade school, men were being drafted. Now, once more we began to see more young men get into those uniforms, including our neighbors like Lehman and Huston McBroom. I was getting old enough and thinking about life away from the rest of my family. Having always wanted to be a real soldier and serve my country too, bulletins posted at school advertising a special training program for anyone interested made the National Guard become a focus. A few months before I planned to persuade my mother to sign for me, I had the opportunity to enlist. (Mom and Dad only knew that I was going to a military school for about 7 weeks.)

The Tennessee National Guard had a headquarters in Cookeville. The person that was responsible for my enlisting was a schoolteacher named Arthur Thompson, Jr. - a close friend and neighbor on Thompson Ridge. We all called him Junior, and I had learned to trust and respect him. Besides me, Junior recruited my cousin Ed, and neighbors Clyde Nickels and Charles Marlow, along with another couple of young students from our high school. So we became soldiers despite how much Dad despised the military.

National Guard and "Summer Camp"

In July of 1952, I was sworn into the Headquarter Company, 3rd Battalion, 173rd ACR, at 17 years of age. Captain Maylor, the commanding officer who swore us into the United States National Guard, issued us uniforms and field gear. I thought, "Boy isn't this something. We are soldiers now!" I was so proud to wear those used khaki uniforms - even though some of them were badly worn and did not fit properly. (The National Guard doesn't do that anymore.) Little did I know at that time, but when he swore me to defend the United States and the Constitution against all enemies both foreign and domestic, the weight of my involvement in the Cold War started to take hold and grow.

We went to "Summer Camp" (as A.J. called the Guard's annual 15 days of basic training) at Fort McClellan, Alabama. I had no idea it would be hot down there in the red clay and tall pine trees. It was much different from the steep hills and hardwood trees such as red oak, hard maple, poplar, and hickory, that I was used to in Tennessee. We bunked in very simple barracks with a cement floor and only screened-in sides. I loved it and thought it was going to be great to be away from the routine at home of having to get up early and water the stock.

The thing I learned on that first morning was that I hadn't gotten away from a loudmouth waking me up! *"Fall out for roll call, and march to the mess hall for breakfast!"* was what now greeted me each morning instead of, *"Get up, and get those mules harnessed for the field."* As much as I wanted to think about food first thing in the morning, the military had other things that they thought were more important than eating. First thing, we had to line up when we heard the sound of the trumpet and were instructed to salute as the flag was being raised. Then and only then after roll call could we go to breakfast; but boy what a treat that was when compared to what my mother regularly made for us. The military breakfast was bacon, eggs, toast, and mystery meat. I was so accustomed to biscuits and gravy that I kept looking around for them. Although I would have preferred biscuits over bread, since toast wasn't my favorite, I gratefully accepted that first full meal. I had not eaten so well since Uncle Bernice's barn raising! During the barn raising, all of the neighbors came to help out. Aunt Mamie and the women chose to give a hand by making a full breakfast meal. And like fueling us up for the hard work of barn raising, all of us new-to-the-military young men needed to be fed well in preparation for the hard work and training exercises that were coming up next. After breakfast, the drill instructors had us do PT (physical training). Since most of us were greenbacks, straight off the farm, the instructors didn't know how far we could be pushed that first day. No one knows if a person is cut out to be a soldier and can handle demands of military life on the first day - not the instructors or the recruits - but it was up to the instructors to find out.

The number of men around me was too numerous for me to count during those training tests. Seeing who remained after the first cut, I thought to myself that some had to have had been passed by default, or that the standards were lower than I had expected for the military. It turned out that the drill instructors had a keener eye than I did at the time; they could see who had potential whereas I could only judge according to the moment. We spent most of the morning just filling out paperwork. The military wanted to know our family and medical histories and the like.

Most of the afternoon was devoted to watching the instructors demonstrate, and then have us recruits do, different exercise workouts. Later that day, we learned to march and parade. Correctly obeying the sharp commands and being a part of the ranks moving in precise order was truly exhilarating! Examples of the commands include: "Atten-HUT" = stand straight at attention; "Right FACE!" = turn right; "Left FACE!" = turn left; "Column, Left MARCH!" = entire line goes left; "To the Rear, MARCH!" = everyone marches in the opposite direction. It was all so crisp and sharp.

We also learned how to carry our firearms correctly during marching formations - the proper term for this was 'close-order drills'. And then, too, learning other specifics such as, "This is a weapon, *NOT* a gun." What excitement for this young country boy! Being in the hot Alabama sun was most enjoyable, and the 10-minute breaks that we were given each hour sure did keep us from overheating.

I awoke the next morning to the same, *"Fall out for roll call, and march to the mess hall for breakfast!"* However, unlike the previous day when all that I could think about was eating, this morning I was unpleasantly reminded that the Korean War was still going on and American Soldiers, Sailors, and Marines were still fighting and dying.

Although there were some similarities from the first day to the second, there were major differences too. While there had been kind of a thrill in learning on the first day, we recruits soon came to realize that the fun and games were over. Any sense of the Drill Instructors (D.I.'s) being easy-going was gone with this second days' worth of marching exercises; and apparently, so were our hourly breaks. Us recruits could feel the change of attitude, and this made us more self-conscious of our actions. For those who hadn't been cut on the first day, learning those beginning steps was hard, and many of us young soldiers couldn't seem to tell the difference between our left and right feet. Although we were told that once we started marching those on either side would fall into step, from the number of people bumping into one another as the D.I. called out change of direction commands it was hard to believe.

Another difference today was, that during the 'close order drill', we were told to keep cadence (rhythm in our walk) at the same pace as the mules they had pulling a wagon beside us. To my downfall, I happened to be placed closer to the mules and I kept trying to stay in rhythm with the mules, rather than the recruits. At the last second, I would remember that I was supposed to be marching in cadence with the men, and not the mules, and would try to correct - which ended up putting an extra step into my cadence. So many times that day that I lost count, I heard, *"Settle down, Bowe, and stop bouncing!"* from the D.I. As a result, I was one of many that had to do a lot of push-ups as punishment for messing up – another thing that we hadn't had to do the day before.

There were so many new and exciting events and sights happening it was difficult to remember the last few days of training at Ft. McKellen, Al. We were basic recruits but treated as if we were infantry soldiers in the first year's training. There were training lessons started in camp that were to be continued at the armory after we returned home. Areas like camouflage could be taught in the classroom and in the wooded area behind the armory. The instructors invited us young recruits to take a hike in the wooded part of the tank range. What we saw, and had not seen at first because of the camouflage, was an artillery piece covered with a net and branches. This was a short and quick demonstration of the camouflage training we were to receive later back at the home armory.

Always happy when payday came, we learned the proper way to "Report for Pay" to receive our money. We actually had opportunities to go into town on 'liberty' during which some of the guys found girlfriends.

The Accident - 1952

Time passed as Mamie and a now graduated Ed Keaton were working at the shirt factory. Everyone in the household worked to help him purchase a car of his own – a 1950 Ford. Who could have foreseen the outcome of that purchase and the trouble it would bring the Bowe family? Ed helped my brothers and I learn how to drive. Together, we would drive Aunt Mamie to her workplace in Cookeville, then us boys would drive the back streets of town. The three of us (Eddie, Bud and I) were best buddies by that time and practically ran the wheels off that car practicing all the driving maneuvers to gain proficiency so we could get our licenses. Four young boys and a car often spell trouble as testosterone and gasoline can be a dangerously volatile mixture. True to form, we began trying out that Ford in different situations - one time the car even slid on the gravel and slammed into a light pole on Old Bixby Road!

When it came to driving, Pee-Wee certainly reverted to some of his undisciplined ways. His old Nashville habits led the Bowe brothers down the fast and furious road to trouble with him on the night we went to a Lafayette School gathering in the Cole Store Community. There, Ed met Bobby Sullen from the Chestnut Mound Community and they somehow challenged each other to a race or, as some tell it, just a fast drive to where US 70N crosses TN56 northeast of Bixby. To get there, the highway is curvy with no passing except for the last mile of the nearly 6-mile run. Since Sullen had his girlfriend in the car with him, I just thought they were planning to open things up through that final, one-mile straight stretch of the highway to the crossroads.

Now they aired those cars out, with Ed in front and Bobby trying to pass! However when Eddy tried to turn at the road leading to town, Bobby - thinking he was still trying to win the race - tried to cut the turn and, *WHAM*, slammed all of us into a brick post; totaling both autos! As I was in the passenger seat and seatbelts were not mandatory at that time, my face and head went through the windshield and were lacerated big time. My brother Buddy, who was seated in the back behind me, ended up with a broken collar bone, and Ed's head tore the sun visor loose. I exited the auto with my legs shaking so badly I couldn't stand. Both of us Bowe boys went to the hospital where stitches and bandages, a sling, and braces awaited us. While he may have been book-smart, Pee-Wee was definitely a dumb driver! To be sure, he probably had some nasty headaches for weeks after that.

Football and Love Life

After three days, we went back to school. I was a junior in high school and on the football team. We had done well, winning our first two games. Though football was going great, my love life was non-existent at that time but I had a longing to change that. Some of my early attempts at relating with girls hadn't turned out so good. I was clearly a novice and didn't know how to treat a real true girlfriend. I had tried to attract other girls before, but as I found out, I had much to learn.

Whenever I had tried to get close to a girl I liked, I would get too physical like grabbing their hands or giving hugs around the waist or shoulders. The result was total rejection. There was one girl I did like when I was 15 and she was different, but she frightened me by asking me to buy her a dress. (I didn't even have money for a coke or candy and here she wanted a dress.) I thought, "Boy does she want to get serious!" That didn't feel like true relationship to me, however; more like using someone. As a 17-year-old freshman, I had wondered if there was something seriously wrong with me. You see, girls that smiled at other boys would just look the other way when I tried to catch their eye.

Now sometimes I took this more personally than I probably should have, because the facial scars (from the car accident with Pee-Wee) were still healing. So, I was hoping all would be fine when I met one of the prettiest girls from the dorm and she seemed to like me too. Joan lived in the dorm with lots of other girls from no telling where - Nashville (where at least one of her parents lived) to immigrants from Cuba. They probably talked a lot about football players, just as we talked about the cheerleaders, and who is the best looking, and who has the most beautiful body, etc. Our friendship grew close enough that, when she asked to wear my football sweater, I assumed that she would consider herself "spoken for". It was just one of those understandings that you took for granted in those days; a pretty girl catching the eyes of several boys and one of them wins out enough for her to ask to wear his sweater. I thought I had won and was very proud to have her wear my sweater and thought that she felt honored to do so.

Well, the game was in progress and there was a time out, so I looked up in the stands to see if I could spot '*my girl*'. Sure enough, there she was wearing my sweater. But wait, my Joan was not in the cheering section but instead sitting elsewhere very close to another handsome boy! I thought, "What is this?!" What a disappointment; what pain! My trusted friend is sitting in the bleachers with another guy wearing my football sweater. Ugh!

Too soon the game resumed and the people in the stands faded quickly from my focus. I had to play my position. (I was what some call a two-way player as I played the right end position on both offense and defense.) In actuality, I was extremely aggressive on defense. Ours was a semi-private, Methodist-Episcopal, small-town country school with both boarding students and village students. The boys' and girls' dormitories usually had between 60 and 70 students. So, though a small school, in sports we excelled such that we were regularly competing against much larger schools and the final score usually made it difficult to recognize which score belonged to which. To be sure, that was probably because some of us country boys were large enough and strong enough to play well on any team.

Joan had a hard time learning the customs pertaining to social relations at Bixby Seminary. She often got into squabbles with other girls. They tried to explain to her that: If you accept something from a boy, then he is more than just your friend and, more importantly - that next level comes with expectations on both sides. The girl stays true to the boy and, if the boy starts paying attention to another girl, his girlfriend has the right to confront him and do her best to find out where she stands in the relationship. We had a good conversation that really made Joan think about the way she had treated me. In the end, we both agreed that we were not a romantic match and she gave me back my football sweater.

Annual training came up in a couple weeks and off I went again. This military training helped my confidence and earned me a few dollars.

National Guard Training

While the Korean War was going on, these men volunteered to serve in hope of obtaining a special job or better assignment than what the Draft Board would have offered. We served together and took special training on the amphibious DUCKS, Jeeps, and Trucks. There were some serious moments for this PFC.

And to think, they were over in Korea doing the same thing!

Tanks are never comfortable to be in; not in extreme cold weather or heat like we had in Alabama—too hot and smells like diesel fuel on the outside and like firecrackers on the inside.

We trained on the (DWKS) or DUCKS as they are most commonly referred to.

A great machine and landing craft that was highly used in WWII and the Korean Conflict.

NCO's were always checking on us young Guardsmen.

In addition to firing on the tank range, we Guardsmen practiced with the M1 Rifle

More Annual Training on the Tank Range - 1953

This was my second "*Summer Camp*" as A.J. Thompson referred to the Guard's annual training. Along with 5 or 6 others from my small hometown, we headed out from Cookeville, TN together for the duration of our training at Ft. McClellan in Gadsden, Alabama. None of the officers or NCOs mentioned the war or that there was a possibility of a call up, but we all knew the unit could be called to active service and deployed to Korea at any time.

Other than the range finder (which was a neat piece of high-tech at the time) none of us "tankers" or men in training on the Sherman tanks from the companies of the 173rd Battalion had been fully familiarized with the tank when back at Headquarters Company. I had been allowed to sit in the driver's seat and view the instruments, but the tank engine at our home base was never started.

For the most part we were still in basic training, well the second phase of it, and we first year recruits had been used around the armory on different projects that had little to do with our MOS (Military Operating Specialty = job title), so we were anxious for real involvement. New replacements always get the short end of the stick and, around day three, we had been placed on the Aggressor Force - dodging infantry and tanks in the fields for what seemed like all day but was actually only about a 3-hour training exercise. (We would have been killed or captured if these events had played out in real time!) That block of training was no fun, but we made the best of it.

As a tank crewman, there is so much to learn about: How it runs on those tracks, moves, and makes the sharp turns, the servicing procedure, the type of armament (weapons), and the different types of shells (ammunition). There is much more to a tank than what is visible to the naked eye; so many small but important details to pay attention to. The outside of the tank smells like diesel fuel because that is the type of engine this Sherman Tank operates on. The inside of the tank smells like spent fireworks or firecrackers. Climbing onto and into this heavily armored steel vehicle is a chore. The trainers or instructors are going to throw as much at us as we can take in during these two weeks of training in the field.

They start first by teaching the proper cleaning process on the 70mm and .30 Caliber. Just like the M1 rifle, the guns on a tank must be cleaned and oiled very lightly. If you do not clean the bore evacuator thoroughly, you may get a face full of nasty smoke after the weapon is fired. (The bore evacuator is a device which removes gasses and airborne residues from the barrel of the weapon by propelling most of them out of the muzzle.) The training sergeant is explaining the role of the team's crew and their locations. During training, every crew member gets to explore each position. The Driver, of course, is supposed to maneuver the vehicle where the Commander instructs him to go in relation to the target or drive to another location on the battlefield. Next the Loader, which is the worst or least desirable job position. He is the hands-on ammo guy who must learn and know the type of round the Commander suggests for the target. The Loader is also in the most dangerous position location-wise because when the breech (where the ammunition is loaded in) goes back after firing, he must get out of the way or it will crush and injure him. The breech takes up a fair amount of the force along with the bore evacuator. The Gunner is the team leader; he supervises the less experienced crew members. The Gunner must also get out of the way of the hot shell casing and be clear of the recoil and breech as it flies open. The Commander selects the targets and gives the coordinates to the other crewmembers, etc. - they are Staff Sergeants and higher in rank depending on the needs of the unit.

The automatic range finders are great! How can we miss these huge targets out on the range in Ft McClellen? A real target, though, would be like a moving deer - difficult to find and shoot. As most hunters know, you must fire before it moves. The lump in your throat gets larger as you zero in and if the squeeze is too hard that may pull off the target just enough to miss. It is fun as long as they are not shooting back! The crew must practice and practice and practice. In the training here at Fort McClellan, we are provided with "look and see" targets (visible to the naked eye). The trainers selected part of two crews to ride in and on the one tank. We rode slowly around the range as the instructor told the crew what to expect. For example, there would be submerged targets whose silhouettes will pop up when the track on the tank crosses a certain cord or cable; and the goal for the crew was for us to try to hit as many targets as possible as we drove past. There are bunkers and gun emplacements up on the hill and the Commander must select the ones that he thinks can be hit within the time frame that is allotted. And finally, there will be another dummy tank straight in front to take out for good measure as the vehicle and crew exit the range and drive out the final 50 meters across the rough road (tank trail). The instructor wishes the crew luck as they move out in their attempt to hit all possible targets.

Finally, after being cramped up inside it for an hour, our crew dismounted the tank so we could stretch our legs. Sargent Jennings, one of the tank commanders, asked many questions to test our knowledge of the information we've covered. It is August and hot - felt like 100 degrees or more inside that tank as we finished the drive around the target range. We Guardsmen soldiers are told we will be going to the pistol range this afternoon and that the 6x6 will be by to take us to chow shortly. Tank crewmembers also carry a .45 caliber sidearm weapon which they need to get accustomed to wearing and using whenever needed.

That was not the most pleasant way to be introduced to the tank, although we had to start somewhere at some time. We hope the newer models will have some air conditioning or cooling fans; however, the Army is not interested in our creature comfort - only how proficient we become with the tank and its weapons platform. There are three types of weapons on these tanks: the 76mm and the .050 caliber, which are mounted on top, and the .030 caliber used by the driver.

In order to have an opportunity to talk and review the training, we from Headquarters Company decided to go to the chow hall together as perhaps some members will have observed more than others and, within the conversations, we will grasp most of what the instructors are teaching us here on the tank range at Ft. McClellan. We'd had some classroom work about the range finder and the use of it inside the tank. The lines on it represented 500-1,000 yards, and we had to know how to convert that into meters and learn when, where, and how and what type to use them on; as well as the grid marks left and right in percentage or yards. Another topic of conversation, as we unbuttoned our shirts to examine our bruised shoulders, covered the recoil of the M1 Grand rifle which kicks like a mule! The M1 Carbine was different in the fact that it was much lighter to carry and handle with more ammo in the clip, but less range and knock-down power. The Marines favored the M1 because of power and accuracy at longer range.

As often occurs in the military training programs, a *"change in plans"* is announced: Tomorrow we will go to the target range in the cool of the morning and then possibly the pistol range in the afternoon heat. The truck will be by shortly after breakfast and pick up the four of us Headquarters guys. One must be sure to drink plenty of liquids for that old 'hot box'. In the chow hall, as the four of us sat joking about school, girls, football, and how others had graduated the year before but still had not found a suitable job, we heard we were going to use live ammo - "Holy Smokes!" Eight or ten of us men loaded into the truck, all of us were excited about using live ammo on the tank range for the first time. This was a joyous group already acquainted with each other and we liked to make fun out of work.

The "Hot Box"

Arriving at the target area where the tanks were already loaded and in place for action - target action that is - there was only one tank in place for the stationary target we were to destroy in this block of training. (Both the moving tanks and targets would come in another block of training.) There were three complete tank crews in training and the idea was to teach us to be drivers, loaders, gunners, and commanders. As we moved along, some of the young 17- and 18-year-olds had not settled down and continued to jostle and pick at each other until Sgt. Jennings stopped the instruction and said, *"Men, this is serious business and we need to go over some safety rules again before anyone fires this large weapon."* He continued to warn us that we could end up becoming a statistic instead of a soldier, saying, *"The fact is, almost as many men are killed and maimed around tanks than are killed on the battlefield."* We arrived at the first tank position parked in a dozed-out bunker type of place.

After one crew had gone through the dry run action and then fired on the target on the opposite hillside, I did not see or my view was obstructed as to where the 76mm round actually went. I assumed it had hit the target otherwise there would have been some instructions to the crew as to why they had missed. Now the crew I was assigned to was mounting the tank and getting into position. We went through a dry run before Sgt. Moss, our gunner and instructor, said, *"Sgt. Jennings, this crew is ready to search and seek and kill the target."*

This block of training was significant for me. More than the excitement of firing a big gun, this time I got to practice being the Commander! While Driver is to be my normal position, today it's my chance to give the directions and coordinates to hit the targets which are some distance across open spaces and terrain. Sgt. Jennings ordered me to pick up the binoculars, find the target, and give the crew what I believed to be the right coordinates. After navigating us into position, I directed the Gunner to line up our sights on a large white target over on the opposite hill. I estimated the target to be 500 yards away or further. "This is going to be good," I thought to myself as I used the rangefinder scope and double-checked with binoculars to make sure that what I saw was the right target. Then I gave the Loader the command to prepare to load and specified the correct type of ammo. Giving the Gunner the coordinates and having him confirm we were on the right target, my next command to him was to make sure his ear plugs were in tight. After that, I reminded him to stay clear of the hot casing and get clear of the breech which comes back like a thunderbolt. The Gunner repeated target confirmation and the round was loaded. I then gave the order, *"Fire for effect!"* Although the recoil nearly knocked me from my stance, I kept my eyes on the large white sign on the opposite hill. Yahoo, that round appeared to hit center target! I grabbed the binoculars to confirm a large hole in the middle of it. The rest of the crew were so pleased that they also began cheering. Sgt. Jennings congratulated me for commanding an almost perfect hit. Making that shot reminded me of the Hot War going on in Korea. Just to think, tanks were actually being driven, loaded, and the gunners firing upon targets at other commander's orders right now! (Much later, I received an Expert Gunner award and a medal for my uniform in recognition for this.) Thinking back, I wondered how anyone sitting still with a range finder could miss such a large target, but apparently some did - even with an experienced gunner.

Now it was my turn to be the Loader for the next soldier in training's attempt to qualify and I remained in that position until one more man tried to make his mark as a go or no-go, hopefully expert marksman in tank gunnery. It would not surprise anyone if I proclaimed deafness and had no sense of smell for the next two meals! Having done my duty at this site on the stationary range, I still had to wait for the rest of the crews to finish out their target practice training and then do a clean-up around the tank before riding over to the next hillside to load up the pieces of the "enemy" (better known as the wood slats of the target we had destroyed). Finally, we loaded onto the Deuce and headed back to the company area for chow and some well-deserved rest after climbing all over the inside and outside of that hot box.

Once we've eaten, changed clothes, and had hour of rest down here in the hot red clay and tall pine trees of Alabama where the temperature seems to climb by the hour, it's usually possible to fry an egg on the body of the tank by 2 P.M. The next block of training will be with the same guys from Tennessee and the 173rd Armed Cavalry Battalion. Some of us, of course, will learn to like and appreciate one another in spite of the differences we have discovered through conversations and teasing; nothing terrible, just simply different. We have each shared space and time sitting or standing side-by-side in the tank or on the truck and as we ate those WWII leftover, out-of-date rations. The permanent cadre don't eat this stuff. (I say "stuff" but some of the other kids have another name for the stuff they are feeding us = body waste!)

Okay, the Deuce is waiting for us to hop on board to take us to the moving target range for our next block of training. The target is a small box car on wheels with a bullseye on the side that moves north and south on train tracks while we sit in a bunker or 'pill box' surrounded by sandbags and fire the .050 caliber at it as it passes by. Each man is given a belt of live ammo to spend firing at this target as it moves at different speeds. This was not much of a challenge for me. The people who run and direct the activity at this range stop the car long enough to change the canvas covering with the bullseye on it before restarting the electric motor that pulls the target back and forth in between each soldier trying to get enough hits to qualify as a tank gunner. Then, we were removed from the bunker and directed to fire from on top of a tank turret with the .050 again. There was enough power there to cause the tank to shudder and shake somewhat - which also made it more fun. We were issued a shorter belt this time and most were glad. While we were having fun, many would have just as soon been home enjoying a family picnic. I was glad to get off that hot box and back on the ground. The target NCO gave me the thumbs up indicating that I had made enough hits to qualify although I knew there were plenty of misses. As every soldier had a chance to fire at the new enemy on the tracks and wheels, and hopefully all qualified, I was doubting if anyone could hear themselves think. It had been one more day of thunder Boom! Cotta! Boom! Boom! We could only understand arm movement directions and deaf-mute language as the NCO's directed us (in our once again sweat-soaked uniforms) to the first aid tent for water and salt tablets.

Weapons Qualification

Next up will be the pistol range. Sgt. Jennings said we must qualify with the tanker's weapon – the .045 Colt. He had told us earlier, *"Become accustomed to wearing and using it as it may save your life and perhaps that of your friend and fellow tanker."* We may be tired and hungry, but the training must go on as scheduled. Since it was too hot for us to be marching the quarter of a mile to the range, he called for the truck to come by the first aid tent to pick us up. Jennings has been a great instructor for this training, he has a super personality and seems to be able to turn his anger at us into some unexpected instruction or a humorous story.

After reaching the pistol range and signing for 40 rounds of ammo to practice with, there were several silhouettes there in a line and the lot of us started firing once the command was given to do so by the range instructor. We fired those .045 service pistols and discovered just how difficult it was to hit a man-sized target at thirty yards. The force of the recoil on the handgun surprised me as the weapon almost jumped out of my hand like trying to hold onto a young calf back on the farm! Wow, I emptied the clip and hadn't hit anything; the rounds kept going into the ground in front of the target! I loaded the next clip and had still only hit the bottom of the target with my next two shots when the pistol range instructor observed my troubles and came to my rescue. He advised me to hold the weapon with both hands and aim for the top of the target. After that, I was finally hitting the target but unfortunately wouldn't have enough bullets remaining to qualify. Deciding not to give me any extra ammo, he instead used his lead pencil to mark up my target sheet as he did for others who couldn't put enough rounds into the target to meet the Army standard. This did not seem to be the correct thing for a good soldier to do but was apparently a common practice to keep the training going smooth or smoother.

Putting the .045 pistol aside, we began our practice with the .045 sub-machine gun, commonly known as 'The Grease Gun'. This weapon is dangerous and has a loose sling and a sliding steel or aluminum bar stock. It is another ideal weapon for the tank crew because M1's, with their long barrel, were too difficult to manage inside the cramped spaces of a tank. (The paratroopers also used it, carrying it with them when they parachuted behind enemy lines during WWII, because the sliding stock could shorten the weapon by 6" or more and the steel frame stock could be completely removed to fire it like a pistol-grip handgun.) The downside of the Grease Gun was the very short range, perhaps longer than the pistol though. Knockdown power with the .045 is what made it powerful, and the instructor reminded us of that fact as we checked out our ammo by stating, *"Always keep the weapon pointed down range."* Again, we moved from the semi-formation (football huddle) to the range target line and began firing at the silhouettes some 20-30 yards downrange. Once again, most of our rounds dipped in front of the targets. Over the loudspeaker Jennings was shouting, *"Aim at the very top of the target, men!"* In the final analysis, they had to use the lead pencil to qualify over 75% of us tankers. Tankers are typically poor shoots but good strong climbers and fighters.

More Tank Training

As we continued our training, we are awakened by the trumpet call known as 'Rev-a-lee' at 5:00 a.m. After a quick breakfast, we are headed to do the stationary and pop-up targets. Part of the course had been explained and demonstrated during our first days of classroom instruction and inside a dummy-tank trainer. This is the toughest exercise of the tank training; however it was fun and exciting as we got to ride on the top of the tank with the gun mounted on the turret and fire the .050 caliber into the targets as the tank travels over the rough terrain. Attempting to keep on target becomes quite tricky as you have to follow it through the gunsight while moving through the bounces and side-shifts of the tank.

We begin with three crews and tanks on this huge, horseshoe-shaped range. When the tracer from the fifty of the first tank sparked the grass and started a fire up on the hill near the bunkers, the second tank was ordered to stop and dismount to help extinguish it. The dangerous, and what could have been fatal, part was that the crew in the third tank had not been informed of exactly what was transpiring, and the PFC kept firing the 0.50 with people among the bunkers on the hill! The tank range director finally stopped the training and had all tankers aid in the attempt to extinguish the blaze. Before the training could resume, the Safety NCO called for one of the company commanders to report to the range and review what had taken place. There would be an investigation about the fire and the lack of communication between the trainer and the men on the tanks, and the fact that the grass should have been mowed and/or sprayed before we approached the range.

When training resumed, it seemed reasonable that tank number three would continue to pass tank number two, and that tank number one would need to restart from the beginning of the range to fire at the targets on each side of the trail and then the bunkers on the hill. I was in the top turret of the second tank as we began moving around the range, trying to make enough hits to qualify as a tank gunner. After firing on the bunkers as the vehicle made the 90-degree turn, the loader who was feeding the belt of ammo through was laughing at the dust and dirt the bullets were kicking up on the hill. Soon after the turn we were ordered to "batten down the hatches" and fire the thirty calibers from inside the tank as it again moved slowly through enemy territory on this simulated run. Again, the targets would be on both sides of the horseshoe-shaped trail and pop up in the middle of it as well.

After making the final turn toward home base, we were stopped and ordered to open the hatches. Told to climb up again and change loaders and gunners for the final run, we would have abandoned vehicles on each side of the trail to hit and destroy if possible. Now I was the loader for PFC Burns from Monterey, and we kept drilling those stationary and the occasional pop-up targets. Passing by the final target, Jim fired as the turret swung around to the rear of the tank and made that dummy tank explode with that last shot! What fun and excitement we had experienced; but we were so glad that no one was shooting back as the conflict was continuing in Korea for real.

Once the CO's and NCOs reviewed our work and scored our target sheets, my scores for the official records were:

.045 Caliber: Marksman – 226 points

Tank Gunnery: Expert – 413 points

Bus Ride to Ft. Jackson - 1953

What an opportunity this was to be a soldier and earn the very first paycheck of my life - not very much, but a check. I was a PFC (Private First Class) E-2 (Enlisted Paygrade 2). Two weeks of sound tank training at Ft. McClellan had me feeling like I was a real soldier now and, upon arriving home, I immediately volunteered for another 5-6 weeks of Duck and Special Mechanical Operations instruction at Ft. Jackson. I loaded the bus for South Carolina with thoughts of the high adventure that lay ahead of me there where I would take my specialized job training.

A couple of interesting things happened on the bus as we were crossing the steep, crooked Appalachian Mountains in North Carolina. One was a conversation with a Native American on board from Oklahoma. We got into a discussion about the Grand Old Opry, which he referred to as "The Nashville Ball". Connecting to a fellow U.S. citizen somehow made home a little cozier. Another thing I recall was how we soldiers were joking around, punching each other, trying to keep each other awake. At one point, I was crawling up the aisle to mess with one of the guys and the bus driver must have thought I was heading up to cause trouble with him. He took out his pocketknife and flashed it back and forth beside his seat to show me I had better not try anything. I suppose we should have known better than to have been fooling around like that. [A history side note that had nothing to do with the career decision, or my future: The Korean armistice was signed in August while I was there in school.]

Fort Jackson South Carolina: NG Maintenance School Job Training

When I first arrived at Ft. Jackson, South Carolina, a Sergeant met me at the gate and asked, *"Have you had anything to eat?"* I thought that was truly kind and considerate since I had been several hours on the bus without food or water. The Cookeville Armory had not advanced me any pay; so, while I had orders, it was more like traveling as a tourist I suppose. The Sergeant walked me down to the chow hall and talked with me as we ate. He looked at my orders and walked me over to the bulletin board before taking me to the master quarters for my bedding and bunk assignment. It seemed like a good idea to scout around the barracks and admire those tall pine trees so reminiscent of the ones in Ft. McClellan, Alabama. Then, going back to the bulletin board, I searched for the school building number and viewed the map of the base to locate where the classes were going to take place the following morning at 0700.

Many of us arrived early to have the first view of the classroom layout and garner any other advance information that could be helpful. Most of the class had arrived and were milling or wandering around the room. A Corporal with the class roster walked in and gave it to the NCO, who in turn said in a mild voice, *"Ten Hut! Soldiers take your seats, please, and hold the noise down while I call your names. Please answer with, 'Here Sergeant' or 'Here Sir'."*

By the time half the names had been called, the captain in charge of the maintenance school walked into the room with three other class instructors and we were again called to attention. All were introduced with proper titles and the area or part of instruction that was their specialty. Then he spoke briefly about the success of the school and shared some positive comments from former students. He also outlined some of his expectations for the class; saying that he hoped he would not have to be involved in the class situations until graduation. We were encouraged to study hard and give the instructors our full, undivided attention. The class was called to attention once again and the captain left the room.

Sergeant Simpson introduced the first subject of study: The gasoline, piston-driven engine and its subparts. He informed us that the power source for the 4x4 vehicle known as a jeep is not a motor but a piston engine. The internal functions of the gasoline-driven engine are simple. First, we will take the engine block that houses the bearings, crankshaft, and camshafts where the connecting rods and their bearings on one end and the piston on the other end were connected by rings to wrist pins. All the component parts have a function. There was the spark plug, which supplies the spark that ignites the air/fuel mixture, thereby causing combustion. The valves, both the intake and exhaust valves, open and close at the proper time to allow the air/fuel mixture in and the exhaust out. The piston is a cylinder-shaped piece of metal that moves up and down inside the cylinder. Piston rings provide a sliding seal between the outer edge and the piston's inner edge.

Summary of the parts of an engine: The block houses the pistons; pistons are the parts that move up and down. Pistons have connections, rods, and wrist pins. The crankshaft is designed to convert the linear (up and down) motion of the pistons into rotational motion. Engine heads must have openings for airflow and holes for cylinders. Valves control the inlet and outflow of exhaust air that goes into and out of the engine cylinders. A camshaft is a rod with several cams profiled along the shaft length that turns to open up the exhaust valves. The cam presses the valve stem up but needs the assistance of the valve spring, bucket heads, and tappet assembly to be returned to their original position. Timing belts provide the action that regulates the movement of the valves. The gaskets are seals to prevent the water and oil from entering the cylinder. The flywheel and transmission are the main drive train including the drive shaft. Together with the component parts of the suspension and steering, the rack and pinion, the axles, and the rubber insulation bushings all work together to reduce bumps and vibrations on the road.

DUCK Training

The barracks (where the soldiers stay and sleep) for our school had not been put in order. Perhaps this was because the Army National Guard's school instructors are different from the permanent cadre or staff at Fort Jackson. For whatever reason, they needed some bunks replaced and rearranged so the Non-Commissioned Officer (NCO) in charge of the barracks collected a list of names. Then, he had us fall into formation to specify some rules that needed to be carefully followed: Loud music will not be tolerated after a certain time at night and lights go out at 10 PM - no exceptions. Officer Willard then lined us up in alphabetical order. "*We need a Barracks Chief for each barracks. Do we have a PFC for Barracks A? PFC Tillman, that will be you. And for Barracks B we have a PFC Bowe, you are the chief for that one. You barrack's chiefs have the special rooms at the end of the building. You are responsible for keeping some order in the barracks.*" We soldiers were then assigned to a bunk (if we hadn't already chosen one to be near a buddy or friend) and deposited our study materials on our beds.

He continued, *"The classes for today will be held out in the field and there will be trucks ready to load you up in the parking lot across the street. When I give the order to fall out, each of you go to your bunks and get the note pads and manuals you were given yesterday. You will be going to the lake today; it is too far to march and some of you may get lost if we were only to give you directions to it."*

At the command to fall out, we hurried to gather the necessary items and get to the awaiting truck. Upon arriving at the training site, another NCO had us get into formation once again and called roll with the names that had been supplied to him. *"Men and ladies, we're going to make a study of a water/land vehicle today and we may take a ride across the lake later this afternoon. For now, when you fall out take a seat in those nearby bleachers and Private Owens will bring a "DUCK" around for us to study and take notes on."*

Now we began our study of the truck-style GMC CCKW 2½ ton, 6x6 known as the 'Jimmy' or Deuce and a half. This one was a newer version that is going to be outfitted to go underwater - well, through deep water. The pond they chose to drive through only covered the hood part of the cab. We were here to learn how to outfit some of these trucks to perform unusual duties: go up unusually steep terrain, climb and ride over trees, and pull unusual loads. They demonstrated the use of the winch attached to the vehicle that was used to aid in these procedures. A winch is a hauling or lifting device consisting of a rope, steel cable or chain wound around a horizontal rotating drum, turned by a motor or crank.

The CCKW came in many variants; some with open cabs, long wheelbases, and a score of specialized models. The version they were demonstrating had a special breather and exhaust pipes (two on each side) that extended above the cab about 3 feet and, of course, some special watertight seals. The truck also had a supercharger beside the diesel motor in order to allow it to burn regular gasoline as well as diesel. We had to study manuals about required inspections; servicing and maintenance; and repairing, replacing, adjusting, and testing the wheeled vehicle's material handling equipment systems. Servicing the auto's electrical system, including the wiring harness, starting and charging, were handled by specially trained technicians.

The 2-1/2 ton modified CCKW or 'DUCK'

These 6-wheel drive vehicles were an amphibious modification of the 2½ ton CCKW trucks. DUKW's (known as Duck's) were used by the military during WWII and in Korea for transporting goods and troops over land and water. Excellent at crossing water and approaching beaches for amphibious assaults, the program was intended only to last the duration of the hostilities and meet the demands of the war. The Ducks are powered by the 91 horsepower GMC Model 270 engine. They were straight-six engines with a 5-speed overdrive transmission and had a transfer case for propellers, plus a 2-speed transfer case drive for the axles.

The propellers and front axles were selectable. From their transfer case, a power take-off on the transmission drove an air compressor and winch. It weighed 13,000 pounds and could run 50 miles per hour on land and 6.3 in the water. The vehicle was 31 feet long, 8 feet wide, and 7 feet high with a folding canvas top. A high-capacity bilge pump kept it afloat. The Duck was the first vehicle to allow the driver to vary the tire pressure from inside the cab. The tires could be fully inflated for hard surfaces such as roads, and less inflated for softer surfaces such as sandy beaches – thereby adding to its versatility as an amphibious vehicle. The DUKW could carry a payload of 5,000 lbs. and 24 troops and had an Aimer mounted ring for a 50-caliber weapon.

After all the notetaking and listening to three different lectures on how to maintain the engine, change the spark plugs, how to adjust the timing and the carburetor, perform various other service requirements, and of course learn how to fill out paperwork like the inspection and maintenance repair sheets, it was lunch time. We had to load back into the truck for that 15-minute drive to the chow hall. Lunch was great, they served fine food in South Carolina - although some complained.

As promised earlier, we did get to go back to the lake for that Duck ride. We loaded up and the NCO even allowed some of the students to practice driving it across the small lake. It is easy to understand why the military had GMC manufacture so many of these amphibious/land vehicles. They were highly utilized by all of the Allies: 2000 were sent to England and 500 shipped to several other countries including Russia; although the Russians subsequently designed and built their own version of it. The water was so pleasant that I hoped there would be room in the swimming pool for me after duty hours.

Upon arrival back at the barracks, I read the message board for tomorrow's activity and saw: PFC Bowe is scheduled for KP at 0400, 12 August 1953. "Well," I thought, "may as well go swimming." They had a genuinely nice pool with too much chlorine in the water. Once in the water, my thoughts went back home to memories of my brothers out in the hot, dry field chopping out the weeds and grasses that could hinder the young tobacco and corn plants from growing. I did miss them, and the rest of the family, including my sisters - especially 2-year-old little Shirley. My mom had commented in a letter that Shirley would point to the place at the eating table where I usually sat and make some gestures as if to say where is he?

Fort Jackson K.P. and Trouble in the Barracks

Pulling K. P. never hurt anyone, but 4 A.M. is a might early for a country boy - even milking the cows and feeding the chickens doesn't happen that early. All the soldiers need to be fed before sun-up even in peace time. The word on the Cold War in Korea is encouraging. News of negotiations going on is better than more shooting and blowing up a country that is already a disaster.

When they brought three sacks of potatoes and sat them down in front of me, it looked like an all-day job. The head cook showed me how to use the electric potato peeler, which is quite simple and handy. It spins and shakes, making a great deal of noise, and water goes through it to wash out the exceedingly small pieces of potatoes and peelings. I hadn't run more than a third of one bag through the peeler when one of the cooks came and told me he had another job for me. While following him, another cook said, *"And so he takes care of your mess? You don't really need him. You need to clean up your own mess, you were just careless."* To which he replied, *"Oh, shut up and go back to your daddy in the office."* I guess people are people everywhere. In my opinion, most of these men were bitter after being drafted and having their lives changed so drastically.

The Mess Sergeant wanted me to stop doing potatoes and clean up after one of the other cooks who had spilled flour or pancake mix on the floor. These guys seemed to be constantly aggravated with each other. They were draftees who apparently hated being there, and they sure did not seem to like me being there - except when they dropped something or needed a hand lifting those frozen meats coming off the supply trucks. This kitchen was like a war zone, every cook taking shots at each other, and I was caught in the middle. They, especially the Mess Sergeant, looked at me like I just came in out of the hog pen!

Before I could get back to my potato peeling job, there was another spill in the kitchen floor; the second cook had a special recipe boil over on the stove. These guys had me confused about my job. When I helped one, the other would get angry with me and at that other cook who had asked for help. Just when I thought the situation could not get any worse, they started throwing items at each other. Throwing verbal barbs at one another was a daily routine but throwing things like cups and glasses seemed very gross and hazardous to me. Eventually, the Mess Sergeant had to step in and threaten the two cooks with punishment.

All of the cooks and kitchen police eat before the main hungry bunch approaches - some had been waiting in line for an hour. It is after 0800 hours and the kitchen is going to get an extra-duty man or two. These men are serving out their punishment for some sort of an infraction such as being late for formation and roll call. The Mess Sergeant put one of them with me and said, *"We need those potatoes. One of you peel by hand and the other feed the potato peeler."* After instructing the soldier on how to feed the machine, I started peeling by hand. We were doing so well for a few minutes and then there was another mess to clean up in the kitchen. This time it is liquid in

the floor and needs mopping, so now they decide to have the extra-duty man mop; and for some unknown reason had me moving tables and light machinery around.

Since we report for kitchen duty early, the KP's also get off early. The soldiers had returned to the barracks after class and were supposed to be resting, but instead two were fighting. Having already hit each other a couple of times, they were now moving around like boxers or wrestlers, each looking for an advantage to make another strike. Someone said, *"Here is the Barracks Chief, he needs to do something."* While stepping between them, I received a push from one. I told them to settle down or you're going to earn yourselves some extra duty. One of the causes of friction between me and the others had to do with the fact that practically all (probably 98%) of the other soldiers I was training with had been drafted. This meant they have had basic training and, after this job training, most of them have orders to go to Korea or Japan for the next two years or more of their enlistment. So, while they had 'US' on the front end of their service numbers and dog-tags, the NG stamped on mine drew a lot of attention and resentment there at Fort Jackson.

A Hard-won Friend

On another occasion while I was still Barracks Chief, a Private by the name of Tony Noland was having fun punching and shoving, flipping covers off, etc. before the class sessions. This activity carried over into the classroom and the instructor cautioned him about the class rules and told him he was in violation of them. On account of this incident, I had developed a dislike for this soldier and his activities. Although I had a printed roster of all names in my section of the barracks, Noland was the only name I remembered when the Staff Sergeant requested two men for extra kitchen help, so I of course chose him.

Tony strongly objected, which led to some name calling, and a fist fight ensued. I allowed him the first lick and it stunned me so that I was only able to land one punch to his half a dozen. (He was a good fighter – I know because I was good myself.) Now down, a crowd gathered around as I regained myself and my footing once again. As I charged into his reach, he managed a couple of hard blows to my head which was bent down and on my backside. I slid under the dayroom table and stayed down for several seconds as he went off to do the KP.

Of course, a fight such as that has other consequences because someone always informs on the unpopular loudmouths such as Private Noland; and this didn't improve my standing as a person in a position of authority either. Soon the Staff Sergeant was questioning me about the event and blasted me for swearing at the man and then fighting. He said, *"We might have to have a hearing on the matter and that man may receive an article 15 out of this."* The next day Noland came to my room asking that I please not press charges against him. I pointed to the knots on my head and said, *"You certainly pack a punch!"* Realizing that I had taken my authority and position too seriously, I acknowledged that much of the fight was my fault and told him that I wished him no further harm.

Shortly afterwards, the SSgt. came to me and thanked me for not taking the matter any further and told me that he would appoint a new barracks chief, but I could keep my room. As it turned out, I gained another friend in Tony and we all had something to laugh over until the class graduated. Everyone knew me as the NG who was knocked on his butt!

More Special Vehicle Training at Fort Jackson, S.C.

Another fine morning began with the top instructor saying, *"You National Guardsmen and Regular Army mechanics are going to witness some demonstrations today. These ideas are to be taken back to your units and it will surprise your leaders what you saw and learned in this school."* These particular 4x4 trucks that we were to learn to maintain were nice smaller trucks to work with and we were shown how they navigate steep gullies and bluffs that are impossible for even a tank to climb.

We students watched from a distance as these six poles were attached to the underside of the truck. Then the truck's winch was attached to a tree up on top of the bluff. Even though this bluff was taller than the length of the vehicle, the poles underneath are also longer than the truck. We stood in amazement at this feat of engineering as they maneuvered this truck up to the top of the bluff by suspending it from the tree above and repositioning these poles like a ladder on the surface at the top and bottom. Eyes wide, I remarked out loud, *"I will never be able to explain or demonstrate that to the NCO's back at the 173rd."* If they thought they were sending me down here just to be a motor pool mechanic, they were terribly mistaken.

Guard Duty

One of the other duties assigned to us while we were at Fort Jackson was to back up or supplement the Military Police (MP's) to help guard gate entrances, buildings, etc. Second Lt. Clifford Smith had volunteered to be the Officer of the Guard. He seemed to enjoy seeing us soldiers march around in formation after performing the regular weapon and uniform inspections of the security duty personnel. It was his job to record our name, rank, and serial numbers and ask some get-acquainted questions before he explained the duties of the posts we would be assigned to. There were gate entrances where we would back up the MP's as they would check identification and passes (if applicable) at the Main Gate and Gates 1 and 2 on weekends, two posts at the ammunition dumps, and one at the rear door of the headquarters building. Around these last three stations, we were only to report suspicious activity.

We were armed with M1 carbines and twelve rounds of ammunition. (A hot war was in progress in Korea at this time.) There were sixteen of us in this detachment, most of which were older draftees, who were expected to pull 4-hour shifts. Three of us soldiers were what you might call 'kids' including myself and Officer Smith. Determined to impress the others, I had gone to the Base Exchange and bought a pair of airborne trousers with the large cargo pockets on the sides. No doubt I was the best and neatest dressed soldier in the ranks. (This didn't really count for anything except personal pride.)

We had assembled in front of the empty barracks where we were awaiting the time to be assigned our stations. No longer in formation, some were discussing the locations of their posts in relation to where we were standing while others were adjusting the slings on their M1's. As he was required to do, Lt. Smith was checking the 45 pistol he had signed out. The pistol discharged and, as it was pointed in my direction, the bullet struck the shrub hedge very close to my knee – so close that I felt it pass by and looked at the leg of my trousers to make sure it hadn't passed through it. Fortunately for the lieutenant, who was white as a ghost, no harm was done to anyone, but all present were definitely startled. Trying to make sure that no one had been hit with the round, Smith asked *"Is everyone okay?"* while he held the pistol straight up in the air and removed the clip. All eyes were in my direction as I pointed to the broken twigs where the bullet had passed through and proclaimed, *"I haven't been hit, the bullet went into the hedge here!"* Next it was time for our guard unit to march to the first post where they had Pvt. Otto drop out of the formation, salute the officer, and man his position. Then we marched to the next post where Pvt. Sinclair did likewise and on to the third post where I continued the routine and so on until all the stations were manned and secure.

The relief watches came out by jeep as it was quite dark in and around some of the outer posts. Very happy to be relieved of my ammunition bunker watch because the boredom of pacing around in that silent, dark place made me very tired, I fell fast asleep upon arriving at my inviting bunk. Suddenly I was awakened by Lt. Smith asking, *"Are you positive you weren't nicked with that bullet?"* Pulling my leg from the covers and touching the place where the bullet passed by, I said, *"Yes, sir, I have forgotten about that accident."* Seeming satisfied for the time being, it was easy to see that the lieutenant was very worried about being in trouble – as I would be too if my weapon had accidentally discharged. In a very short time, he came back with more questions for me about where the bullet passed in relation to my knee and talked about how foolish it was of him to have the weapon pointed in that direction. I guess he did need someone to talk with besides the Sergeant of the Guard whom I hadn't met yet. (Knowing that a SoG was always stationed to monitor the security on the watch kept us alert at our posts.)

Soon after 8:00 a.m., there was a swarm of activity with a group of officers wanting to question everyone that was present when the incident occurred, including an MP who had heard the shot and had relieved Lt. Smith of his weapon and taken the empty casing with him as they left. Sending out a jeep to pick up him and the others, they asked me where I was standing at the time and where the bullet struck. Telling us to expect more inquiries about the accident later, the Sergeant of the Guard dismissed all of us for morning chow.

Returning to the barracks, knowing we had the day off for rest and free time, we stopped to check the bulletin board for duty details and other vital information and learned that the base will be involved in a parade. Wondering if it was to celebrate the armistice signed in Korea, we hope the news is correct because the rumors are that the members of this vehicle maintenance class are headed to Japan for more training.

Parade preparations include shining our shoes and shaking the wrinkles and folds out of our Class A uniforms as all personnel will be inspected by MSgt. Mullins, the top NCO of the school. Mullins said, *"You folks have done well in school; only a few incidents occurred that we had to address. For the most part, this was a successful, uneventful training session."* As it turned out, this was actually a graduation ceremony for the class.

Back to My Hometown and School Autumn 1953

After graduation from Maintenance Training in August, I went back home where everyone was glad to see me – especially my little sister, Shirley, and my dad who needed some work done on the farm. School began and I was a starter on the first team playing on both sides of the ball, defense and offense. This would be my last season practicing and playing high school football with my two brothers because, after this, I would age-out at nineteen and no longer be able to play – not to mention that my grades wouldn't have allowed me to do so anyway. I became aggressive on defense and lettered that year but did not do so well in my position as right end on offense because I was poor at catching. Our football coach informed the girls in my algebra class that I was a star on the football field. Of course, that was more to build me up as an athlete and player than to get me a date or some girl's interest, but my romantic life did start to improve within the next month.

Not long after I returned from Fort Jackson, the National Guard was called up to fight a forest fire in October near the small towns of Jonesboro and Clinton, which are closer to Knoxville than to Gatlinburg but still considered to be in the Smokies, as the mountain range in east Tennessee is called. I, for one, was quickly packed and ready to go to help extinguish that fire. (Or was I just excited to put on that uniform and go visit a new place and earn a few dollars?) None of the sergeants or squad leaders questioned me about being at the armory to help load the trucks that were going with the unit; they were glad for another able body to help with the equipment.

On the second day I was high on the mountain, battling smoke and small flames with what looked like a garden hoe, when the Company Commander, Captain Maylor, came alongside me and began a conversation while we worked. As we talked, he recognized that I was one of his newest recruits and asked a few questions (some of which I did not have a ready answer for) such as, *"Who is your supervisor and NCO?"* I had no answer and could not identify the person that directed me into the area to extinguish the flames. Another one was, *"Aren't you supposed to be in school?"* I had an answer for that, *"Yes Sir."* He seemed both angry because I wasn't supposed to be there due to my student status and pleased that I chose to be there anyway. I was in a catch-22 situation. Captain Maylor smiled and said, *"Get your butt off this mountain and back in school Private Bowe."* That day and perhaps a few days later, I learned where the priorities were even if I was failing my subjects - school.

Beverly

Shortly after that, a girl called Beverly Johnson caught my full attention and quickly became my very first girlfriend. She understood the school dating protocols. Beverly and I talked in the hallways in view of all the students, then we began to pass notes and greet each other in the morning before classes. As the friendship grew, we exchanged information on our class schedules. Frequently, she would take my hands and push the cuticles back on my fingernails as we sat out on the lawn near the water fountain. Although I really didn't care for this, I decided it was worth the attention to tolerate the treatment. It was such a great feeling just to know that some other person really cared about me. Previously, I had felt like a mess; almost impossible for anyone to like. We Bowe kids were like the wild bunch, undisciplined for the most part. I would have to say that Beverly had a wild side also.

The scars on my face and forehead from the car crash were ugly during healing. Some of the teachers (Mrs. Cunningham for one) tried to get me to rub Vaseline on them; saying that it would make them less noticeable. A very significant memory from that time is that I don't remember Bev ever making any remarks or comments about my scars - that really helped my self-esteem. She was a girl who seemed to be very compassionate; always expressing sympathy over sick animals. The medical bills were coming due and, of course, Dad and Uncle Bernice were very stressed because they didn't have the kind of money the hospital charged. Our only option was to go to court and sue Bobbie Sullen. He was at fault by law and his driving mistake had caused a lot of misery.

We initiated a court case against Bobby, suing him for damages. Ed's car was totaled and had to be replaced and our doctor and hospital bills had to be paid. We settled out of court. Bud and I were awarded some cash money for our suffering (less than a thousand dollars each as I remember it). Dad, being the kind of a character that he was, felt that he was entitled to at least some of the money because both of us boys were still living under his roof. After many arguments, discussions, and negotiations about the money, we knew we would never overrule our dad so we just succumbed to his will. Since his word was law in our home and family, we finally came to an agreement that I would buy the 1939 Chevrolet that dad used to transport his hunting dogs in. Although I was forced to pay his price instead of a more reasonable, actual value for the vehicle, his kennel-smelling vehicle did run after some attention.

Two things about that old car come to mind that I shall never forget. First, old as it was, this vehicle provided a chance for Bev and me to go on a date and get away from school. Second, this old car had an on again off again electrical short somewhere in the wiring. At times when you least expected, the old car would not start or the lights wouldn't come on. Dad taught me the work around of how to jiggle some of the wiring and temporarily fix the problem. I had not yet learned about electricity or much about autos in general.

1953 was such an interesting year:

The football team, the auto accident, joining the National Guard, and meeting and dating Beverly Johnson (Miss Detroit).

This picture is only an image of Miss Detroit. Our relationship only lasted around two years; about a year after I entered in the Air Force. Most people remember their first love; the fond memories and perhaps heartbreak which serves to make us grow as a person.

Bixby Seminary High School

TV was new and brought the news into our living rooms. The Korean War was going on and we were exposed to some news on the Huntly and Brinkley evening report about real shootings and body bags seen in tarps lying on the ground.

While Miss Detroit was in the boarding school, she found a way to go out and visit her kin in the county nearby. I managed to drive the 1939 Chevrolet and pick her up at the place she directed me to—a location some ten miles away. She was dressed up so beautifully, but was overly so for the ride in the hound dog smelling Chevy. Nevertheless, we didn't mind as it was the most happy day we spent together.

Pee-Wee's 1952 Ford before our accident

The auto accident hindered my football season and left permanent scars on my face, but may have protected me from other unseen things.

Bobby Sullivan's car after accident

D. Ray Bowe —1953

Sophomore 1952

A3/C Bowe Jun. 54

The Lifetime Memory Date

Beverly Johnson was a city girl from Detroit, going to school down in what we call "the sticks" - the hills, valleys, and ridges of Tennessee covered with briars, bushes, and trees. Her parents still lived in Detroit (though not together) but she did have some relatives living nearby in Gainsborough – a community in Jackson County, only a few miles away from Bixby. How Bev arranged for permission to go visit them is beyond me. Believe it or not, I was 19 years old when I went on this, my first official date. No doubt, Bev drew a street map which gave the address and description of the house to me in one of the notes we passed to each other in the hallway. I can visualize that house up on a high bank; much like Grandpa Jim Bowe's place off highway US 70N. Beverly was dressed up so gorgeous! Just looking at her was enough to make me fall in love all over again. We drove around talking and laughing for hours, having the time of our lives. I had compassion for all the girls living in those dorms, as they had no life except for school studies and the four-block walk to either the Methodist or Church of Christ there in town. This is what made the date a double treat for Beverly: Getting out and away from school *and* riding with her boyfriend.

Whether at her insistence, or my concerns over the gas money - and perhaps a mistrust of the old Chevrolet - we didn't venture too far from Bixby. Funny, but somehow, I don't recall exactly where we went; maybe the Dipsy-Doodle Drive-in near Cookeville, or a nice restaurant called George's Place at the crossroads, or it could have been the Cumberland Café in town. All of my focus was on enjoying watching and listening to Bev. She was so happy just being out among people and away from school. Struggling with my desire to put my hands on her and pull her close to me, I was tongue tied as to what to say except to tell her how lovely she looked in that beautiful dress and how beautiful her smile was. I felt somewhat like Archie Andrews in the old radio series The Archie Show. He and Veronica were close friends and he never had the words to say to her. She would say nice things to and about him while all he could get out of his mouth was "Gee Whiz, Veronica" and laugh his silly laughs. That is exactly what I did, laugh and laugh as I looked at her beautiful face.

I still have her school picture; and, though most would say she was only average looking, she was a queen in my eyes at the time. Again, the thoughts that ran through my mind as I would look at Bev were, "I would like to be with her the rest of my life." Much later when my buddies in the Air Force would look at her large picture in my room, they would say, *"You have got it bad. You better marry that girl!"* I totally agreed with them on one level; but deep down I also had many serious personal dilly-dallies to deal with. That date was a memorable first, but we had others too.

One of the school groups had a candy counter in the basement situated between the men's lockers and the stairs in the back of the main building. The counter wasn't open every day; some days it opened at 8:30 or 9:00AM. I went down early one morning to find two or three of the football players totally engaged in some heavy-duty smooching in that locker room. I thought, "If those girls are willing, maybe Beverly will go along with some of that." I suspected though, that if we ever got caught, she would be punished and possibly even dismissed from school.

When I talked her into meeting me down there, she was surprised to see who else was there from town. I was the only one who rode in on the bus every morning. What a feeling, to hold my beautiful girlfriend in my arms for the very first time! (Even in a locker room - Wow!) Trouble was, we had to be quick; couples couldn't be in there more than five or six minutes before the bell would ring and everyone, especially the girls from the dorm, had to be in class. We carried on with that surreptitious activity for weeks until the Dean got wind about something going on in the boy's locker room. No doubt our teacher, Miss Hill, had plenty of loyal informers.

I am sure that, had she been paying close attention, our teacher probably would have known who was going with whom. Miss Hill tried to help me and I liked her. I was careful to be polite and very respectful to her, especially in her Algebra class. Once I made a remark to her about disliking something and I'll never forget her reply: *"There are so many things coming your way that you are not going to like young man."*

Anyway, after one smooching session, Beverly and I were headed up the staircase when Miss Hill looked down and said, *"Beverly!"* in that correcting/chastising tone of voice. As Bev replied, *"Yes Ma'am Miss Hill,"* I stopped dead in my tracks wondering if I would be called to the office as well – thankfully, I was not. I don't remember the outcome of that encounter but that put an end to our morning meetings. It became kind of lonely not being able to see and hug and kiss the one I was so much in love with. From Bev's behavior, I can only guess she received some strong lectures and warnings about me; including a reminder that her parents had sent her there to study and learn - not to spend time fooling around with wild football players.

One unforgettable Sunday, Bev and I were out enjoying time together but, knowing that she had to be back to the dorm by a certain hour that night and that all great times must come to an end, we kept close track of the time. I carefully planned the route we would be taking back to school by way of Tom Falls Road, knowing that there was a place where we could pull off the road and into a thicket of trees for some quiet time together (hugging and kissing until our lips were sore). Trying not to mess up her dress we kind of laid over in the car seat together. We were both scared, excited, like two racehorses wanting to go all the way with our love. But we both knew it was wrong. It should not and absolutely must not happen. We must not let it happen. Bev said, *"We must go, we must go."* I agreed, we must go now before things got out of hand.

Well by then it was dark and for some inexplicable reason, that old Chevrolet motor simply would not start! Knowing that I had to get out and fiddle with the wires again but still hesitating because I was dressed in my best clothes and, like any old car, this one was dirty, greasy and oil soaked, I knew I must try to get this car going or Bev would think that I had purposely sabotaged the car in order to keep her there in the dark woods. I raised the hood but, try as I might in the dark, I couldn't locate the correct wires. Time was short; we had to walk. There were no phones in the nearest house and it was a mile or more back to the school and the girl's dorm over dark, rough, rocky roads. Bev was smartly dressed and in heels no less. Recognizing that wouldn't work, she slipped off her shoes and opted to go barefoot. She was a good sport about it and there were no complaints from her - even while walking on rocks for what seemed like a very long mile. We were very concerned what people would say about our late arrival. What would they tell the authorities at school? Upon hearing our story, would they believe us? Would they believe we behaved ourselves, or would they think the worst? Would Beverly be dismissed from school? Boy did I have some explaining to do to the sheriff about the abandoned car also!

In the end, it all worked out well. I caught some teasing about that night because I had to get an experienced mechanic to go with me to retrieve the old Chevrolet. I suppose Beverly was also teased and probably asked all kinds of questions. "Where did you go and what did you guys do?" "Did he let you drive the car?" "Did he kiss you?" "Did he give you a ring or talk about marriage?" I can only imagine some of the hundreds of questions that could have been asked. Perhaps the fact that both Bev and I were not troublemakers and we were trusted and even liked by many of the students, helped make our story more believable. Most knew that I attended church and played football.

I must say this to the whole world and be very clear about it: When measured by all possible high moral standards, Beverly Johnson was a fine young lady. Oh yes, she was spunky and a little bit mischievous - otherwise she wouldn't have had anything to do with me - but I left Beverly Johnson that night just as I found her: pure and clean. It is my hope that she stayed that same way until the day she married.

There were many other good times we had together. One was a ride we took together in the back seat of her father's car – a fine, new model. I think it was over the Christmas holidays that her dad came down from Detroit, his place of business or work. It was easy to see that Beverly was "daddy's girl." He drove her to our house and then we went around the Bowe Ridge down to see the lakes and rivers. Mr. Johnson seemed totally open, happy to respond to every wish of his sweet little daughter. I have often wondered if that delightful ride in a fancy car was superficial or was he the one behind her being in a private school to learn the faith and practice of the Christian religion instead of going to a public school in Detroit?

May 1954 and school is out! I had not done well (flunked algebra and another subject), and it was clear that school did not have my interest. Dad and I were not getting along so I was not doing good at home either. My brothers and I had no leadership at home. With proper counsel I could have taken building trades classes and perhaps graduated with the class of 1955, but our father had no education and thought one was mostly unnecessary. Something had to change! I desperately needed to come up with funds but where and how? The military had begun to open my mind to realities and possibilities that I had not considered before - so many things beyond becoming a poor sharecropper like my father. I cannot remember any reasonable options except going to the Air Force Recruiter's station - especially after an incident that occurred with my father.

It seems dad wanted to go to the stock sale in Cookeville one Friday; and, either was broke or had other purposes for his money, so he wanted *my* six dollars. Having failed to ever do so in the past, I politely said no because I knew he would not pay the money back. Dad's response was, *"You do not live at this house anymore!"* The great God of heaven and earth has a plan for each of his chosen and He makes a way mysteriously. The way He made for me was the United States Air Force.

My first memory verse was Romans 8:28:
"But all things work together for the good of them who love the Lord and are called for His purpose."

I was nervous entering the recruiter's office. He gave me a room in the hotel for the night and roused me early the next morning for breakfast before seeing to it that I boarded the Trailway Bus to Houston Barracks Testing Center in Nashville. Once there I met several recruits and we wandered the streets for a couple hours until the time we were to report in for the evening meal and lectures. We gathered in a circle to smoke and tell our tall tales before the old timers shared some of their fun and encouraging stories.

The older vets gave us an encouraging pep talk.

Houston Barracks, Nashville, TN

After the morning meal, none other than Gene Autry gave a patriotic speech. Then, after more recruiter speeches, we were bussed to Houston Barracks for testing. When asked about my faith, I hesitated to answer because I was confused by the word faith. Remembering faith as a religious practice, I had failed my Lord in actions and repented before answering, "Protestant, non-Catholic."

I rode the train a couple days to Texas and met the DI's who seemed to dislike us recruits. We watched the most recent graduating class march in review and I wondered,

"Wow, Will we look like that? "

Train to Lackland AFB, TX

We soon learned new words like 'Airman' and 'Flight' and a new way to address the Drill Instructors -"Yes Sir, Airman Davis!" or "No Sir, Airman Davis" Always before speaking it was "Yes Sir" or "No Sir".

Air Force - May 1954

My best friend left Bixby first. We said our goodbyes at the bus station where she was headed back home to where her parents lived in Detroit, Michigan. Beverly and I promised to write each other every chance we had to do so. At that moment, we thought we were very much in love at the time; "*Young love filled with emotion*" so goes the song. Did we know the true meaning of love? I know we were certain about our feelings for each other and wanted to be together. We so desperately wanted our relationship to continue and last forever. Many young couples run away and get married – I almost wish we had.

About a week later, my brother Bud and I were sharing our thoughts and plans with one another as we drove in the 1½ ton truck headed for the pasture to milk the cows as we usually did each afternoon. This three-quarter mile journey was often our only time to communicate and discuss the family problems and our concerns about them. Our father had a lifestyle that did not include our mother or us (his children) very much; and he had a nasty temper we all had to deal with. Without any sort of counselor or advisor to help, we knew we had to make plans for ourselves. As he and I talked about our personal needs both short and long term, I mentioned how I didn't even have a suitable change of clothes and my teeth were going to need some expensive dental work soon. Bud was an unnoticeably quiet person and a very good listener, which was extremely helpful at times. No matter how difficult your homelife is or was, it is hard to leave the rest of your family behind when one finally decides to go. (Our home was not much, but it sheltered us from the thunderstorms and the cold winter winds.)

Somewhere in there the decision was made. I was going to join the United States Air Force – a choice that would change my life forever; not just the next 3 years, 7 months, and 21 days in the service. After having served in the Army National Guard there for almost 2 years, getting the 13 miles to Cookeville without a car had not been a problem. Usually a recruiter is happy to see a warm body step into their office and fill out papers of intent, but for some reason they could not swear me in right then. But, to discourage me from changing my mind overnight, he provided a meal and a hotel room for me to stay in. So the next morning, after being treated to breakfast by the Cookeville recruiter, I was on a Trailways bus to Nashville.

[4]

My excitement continued to grow after arriving there, joining up with several other prospective soldiers, and being met by a group of recruiters from the Nashville headquarters. Many of us had meal money but, for those who could not provide for themselves, the military had verbal agreements with restaurants and hotels. When an expense occurs, there is a ticket receipt generated from the business for the recruiters to sign and to list the name of the person who incurred the bill. Hotel room numbers were also issued (two men per room). We were required to listen to the speeches and instructions given, which included orders to report to the hotel lobby the next morning.

When the Air Force told us to write home, I probably included these exciting details. The next morning, we assembled in the lobby to meet with the recruiter: Gene Autry!! He gave a very good patriotic speech, shook our hands, asked our names and what town we were from; that made us all feel special and I was even more excited.

Afterwards, the recruiter put all of us potential recruits on a bus to go to the Houston Barracks in South Nashville for testing. Once the tests and some interviews were finished, it was time for the placement sergeants to give me a choice of what my job was going to be. He wanted me to become a lineman; I suppose because that's what they needed. I stated that I wanted to work on and around the aircraft. While seated at the officer's desk, I had a mind and perhaps a heart change and repented the moment the counselor asked, *"What is your faith?"* The question surprised me and sent shockwaves all throughout my system! It was almost like the Lord was saying, "Where have you been Dorris Ray?" (Like in Genesis 3:9 when God asked, "Where are you, Adam?") A moment later, the officer asked me if I were Catholic or Protestant. After hesitating for a few seconds, I responded, *"Oh Methodist; no, maybe Baptist."* But as the officer wrote the statement down and filled in the necessary blocks on the interview sheet, I was thinking, 'Yes, I am a Christian - even though I have not acted or thought like one for quite some time.' There definitely was a decision made by me that day and I prayed that very night and all the way through most of the nearly 4 years of nights to come - minus a few lapses when some worldly moment or thing or person had me side-tracked. I knew Scripture well enough to recite John 3:16, Psalm 23, Romans 8 & 9 and many of the Bible stories, especially Jesus' Sermon on the Mount.

The bus ride back to the recruiter's hotel in Nashville was quiet as some of the recruits had failed the grueling tests we were given that day. I felt sorry for an African American man who sat beside me on the bus ride back from the test site. Although most of us did pass, some of those young men never made it into the Air Force because they didn't. I can only imagine how defeated those men must have felt.

Back at the hotel, after those being dismissed were sent home, more patriotic speeches and a nice meal awaited those of us who would be taking the train through Arkansas and into Texas the next day. There, the other recruits and I met these mean and very rude sergeants who promised to be our new mothers and fathers. Of course, they were only imitating the Drill Instructors (D. I.'s) we would soon encounter at our new home for the next 12 weeks. All of us learned new words such as 'Airman' and 'Flight' instead of 'soldier' and 'company' or 'squad'. Some men were sent home after reaching basic training; and some were even taken to the brig (military jail) for signing fraudulent enlistment papers.

Another event kept it outwardly apparent that my Christian faith was affecting my lifestyle. While at the hotel, we young recruits sat around in a circle telling stories and jokes as young men so often do. At that time smoking was permitted just about everywhere, so someone lit up his last cigarette and I pulled my pack of Cool's from my pink shirt pocket and lit up. Then, the recruit sitting beside me asked for a cigarette and so did the one next to him. The pack was passed around the entire circle with each helping themselves to one before it came back to me. Observing what was taking place, I decided then and there about smoking and buying cigarettes and told the last guy to keep the rest. That was the last pack that I purchased for myself. Even today, the social habits of smokers still include those who smoke but do not buy; asking their friends for a handout of tobacco and other items. Those recruits who passed all the tests would move on to basic training via a train ride of perhaps two days travel to Lackland Air Force Base in San Antonio, TX.

Air Force Flag

A. F. Basic Squadron

Flight 499 - June 1954

Lackland Air Force Base, TX

Trying to wash and shave with all those other faces and arms in the way was a challenge.

Air Force

Trevino

When recruits stepped off the train and into trouble

D. I. Dumas- A rough man

In Air Force Basic Training we were issued olive green coveralls that fit no one and told, "You are in the Air Force now and not behind a plow." Our mother no longer made our bed and we had to do it correctly or there was trouble.

We marched and marched; our new boots did not fit that well. We Double-time Marched, Route-time Marched, To-the-Rear and Right-Flank Marched, Pin-Wheel Marched, and Column-Right and Left Marched.

Marching in Parade Dress Uniforms

Air Force Boot Camp: Lackland AFB, San Antonio, Texas

After 10 years of listening to the train climb the 6-mile hill up Buffalo Valley and blow its horn for the Silver Point stop from my parent's back porch, I was finally going to ride that train! That first train ride was outstanding; it's hard to believe I hadn't ever done it before. When we boarded, each recruit was issued a seat number and provided a standard meal for free. We could pay for the option to order off the dining car menu but most of their prices were outrageous. All of us riding together were joking and laughing and having a great time until we got off the train and were met at the San Antonio station by three nasty, grouchy sergeants who did not seem to like us very much.

They took all of us to the Air Force Base and we were escorted into a building with signs on the walls with sayings such as: "You are in the Air Force now, not behind a plow." "Your mother will not be cleaning your room or making your bed here." "You need to give your heart to God because we have you're a$$." Other large boards and placards contained signs with rules and various statements. I remember reading one that said something to the effect, "If you curse and use vulgar words, it is because you aren't smart enough to use correct ones." This of course was part of our orientation into the Air Force - it was the start of a completely new lifestyle. Then came listening to people raising their voices as if we all were deaf. I remember the Senior TI (Technical Instructor), Airman First Class Dumas, was our daddy and mommy for the next few weeks. Any word that came out of our mouths had to be, "Yes Sir, Airman Dumas!" and "No Sir, Airman Dumas." We were then marched to another building and issued our clothing and bedding. A footlocker was already in the barracks where we were going to live for the next 12 weeks. I thought that the green coveralls they issued us to work and parade in were the ugliest garments that you could possibly put troops in. The boots were strange to look at, but I liked them.

Our first instructions had to do with grooming "*Air Force style*"; the how-to on making up the bunks properly. There were daily inspections of our double-decker bunks, footlockers, and the area around them. Next, we learned the proper way to clean the latrine (bathroom), the proper way to do barracks security, as well as 'securing the post' by quoting the general orders and answering the question asked by the officer of the day. He would ask, *"What is your duty on this post?"* We answered with our name, rank, serial numbers, and general orders about the post. We learned the correct way to march by staying in step, counting cadence, and doing column movements. We did many more things as a unit before going out to the field for those outdoor exercises such as pitching a tent and bunking out all night. My tent and bunkmate was a very nice Mexican young man named Ramos. He was so much fun; we laughed most of the night instead of sleeping.

Most in Flight 499 (the number of our basic training group) were fresh into the military and all were simply basic airmen, E-1. I had to retrain myself to call our training unit a 'flight' instead of 'company' like it had been in the Guard. Due to my previous military experience, I thought general boot training was going to be a breeze, but my time in the National Guard soon brought me additional responsibilities. The voice of T.I. Dumas's words still rang in my ears, *"Flight 499! Come to attention!"* Airman Dumas came out of the barracks with a handful of clothes on hangers and threw them in front of our formation. *"Disney, Sullivan, and Fleischer!"* he yelled out in his rough, booming bass of a voice. *"Pick them up and be quick!"* he directed the three recruits. *"You people are no longer in my flight. You people cannot file fraudulent papers* ***and*** *be the type of people we want to retain in the Air Force. YOU PEOPLE, follow Airman Shoemaker to the Provost Marshal's Office."*

This must have been especially painful for Airman Dumas because he had personally picked AC/3 Disney to be the flight leader; thinking his age and prior service would be good for the squad. The flight leader position oversaw several helpful functions and was designed to make the training instructor's job smoother. Disney had demonstrated the ability to handle the men sternly and get things done. After calling all of the squad leaders and a short list of other people into his office, he ordered the rest of the flight to leave the area with D.I. Kirkland. It took some time for this event to soak in. There was some sympathy for those three as we had grown accustomed to seeing them every day over the last three weeks. I agreed that it was a sad situation, but on the other hand Disney had already made several enemies. The private session Dumas had with his squad leaders had gotten quiet, which seemed to mean that he had settled down. We were informed that Airman Trevino had been slipping out of the barracks at night to go home to his parents who lived within walking distance of the base. Since he had made roll call every day, it was obvious he had assistance in his deception. *"I have a plan for him and a few others that have been stretching the rules,"* Dumas stated. *"Do any of the rest of you want to let me down and go home?"* Obviously, he was taking the situation personally.

The apparent shift from strict and rough leader to buddy looking for help didn't sit easily with any of us recruits. We hadn't seen this side of him, where he was asking for help, as he had to choose a replacement. I saw him look in my direction and quickly looked away. *"Anyone wants to volunteer,"* he threw out with a laugh. When no one stepped forward, my previous experience got me elected by default and he said, *"You are it, Bowe, for the time being."* To the others he let it be known that, while he did need help and cooperation, he was still going to be hard on us and not to expect any favors or breaks.

I didn't want the position, but I knew that it would be in bad form if I were to turn it down. Almost immediately a conflict arose with my leadership as squad leader when I was trying to instruct. An African American man by the name of Millington expressed resentment and seemed to threaten me, so I backed off. This course of action didn't sit well with T.I. Dumas because he wanted his squad leaders to "crack the whip" over the squads to make his job easier. He called me into his office to inquire; and by inquire that meant that I took major heat for not putting Millington into his place. In my defense, I mentioned how I had grown up protecting my two younger brothers from my dad's abuse so being a hard ass did not fit my DNA. Seeing as how I wouldn't be the squad leader he needed, my position was given to an airman by the name of Rodden Kincade.

While I wasn't happy about falling out of Airman Dumas' good graces, I was glad to be rid of the responsibility and still believed that the Air Force was my career. A few days passed before I asked to see the commanding officer, and was given the opportunity to visit the Wing Command Office for record updates, etc. At this time, I informed them of my experience in the Tennessee National Guard and they in turn changed my pay grade. I was allowed to keep the paygrade, but not my rank of PFC (Private First Class). Another item of business was making out an allotment to my parents; my uncle had advised me to do that before I left. (Bernice was a good counselor and adviser.) Since the pay was more than what I needed to take care of myself, he had encouraged me to send money home to help my family. My mother also set some of this money aside so that I could eventually buy an automobile.

Yes, we learned many new things: how to wear the new uniform, salute, and march. Part of our training also included learning how to clean, fire and use the M1 Carbine Rifle and march while carrying it and a backpack. We also had to run, crawl, and climb the obstacle course. Additionally, we learned how to construct a 2-man tent and live in it if necessary. Next in the training agenda was the 5-mile march, in addition to the obstacle course with field and tear gas attacks. We were to swing across a muddy pond on a rope (some went in the pond of course), then run and climb over the barriers, and crawl through tunnels, etc. The "best of the best" attitude of the high school football player in me came out during the obstacle course. I was in excellent shape and was passing all the slow, overweight people, striving to be in the front of the pack. When I crawled through the dust-filled tunnels, one of the instructors hit me hard on the helmet that I was wearing. In my pride in striving to be the best, I took this action from the instructor as an afront and proceeded to get into his face. I was about to get physical with that medium-sized man when another instructor, observing the situation, stepped in to protect both of us. He nodded his head down the trail and told me *"Get going."* I knew that the best thing was to be obedient and finish the course rather than trying to defend my pride.

I wondered several times if they might delay my graduation by a couple of weeks, since this had been known to happen to five or six others who were not fully cooperative. One of the times that this thought crept into my mind was after a 5-mile march. We were approaching a creek bed where our instructors were awaiting the training group with teargas grenades. I went to put my teargas mask on my face and found it to be leaking. When I tried to get the instructor's attention to get support on how to get my mask to fit more securely, one of the other instructors came over with a baton and hit me with that thing. In my haste to get out of the teargas, I grabbed him and threw him down into the creek bed. When I no longer had to worry about his baton, I ran as fast as I could to get out of that gas-infested environment.

Aside from being hit by the instructors, I thought basic training was so much fun! Being a Drill Instructor may have been a good job for me had I not already made my choice to be a jet mechanic (mechanical technician). It worked out because mechanical design seemed to be in my blood.

The firing range is part of basic training. It certainly did get the higher up's attention when a certain Korean War veteran started informing the target range instructors that they weren't conducting the training properly. His observations closed the training session down for a time while the instructor's weapon experts listened to this seasoned soldier's training perspectives concerning the user end of those M-1 carbines. Because it was shut down for so long, we were given very little weapons training, but we did learn to clean the M-1 Carbine really well and went to the firing range to qualify. There was also a swimming pool where we were taught how to rescue a wounded, floating airman. These instructions included jumping off the high dive and treading water; how to remove our boots while trying to stay afloat; and how to rescue a drowning person; as well as how to do CPR on those that had water in their lungs.

While in Basic, I also regularly went to the base chapel to attend services. I was already in a pattern of reading the Bible and praying each night. Others had many names for me, but most called me "preacher" because, as a new believer, I always knelt beside my bunk each night.

After a grueling 12 weeks, finally graduation day and new orders came. Most of us were bussed to the flight side of Lackland Air Force Base and loaded onto air transports for new destinations. Those 3 months of Basic Training sure went fast; and, at the end of it, each graduate had to sew our stripes onto the sleeves of our uniforms. The first stripe that we received was for our graduation out of Basic. Of course, due to various circumstances not all graduated, which is always the case. Some didn't because of misconduct like being AWOL (Absent Without Leave). Others never made it through the third week of training. We all had learned so much through our period of Air Force basic instruction. With all that good training behind us we thought, "How could we fail our folks and friends back home that were depending on us to defend them?" Now it was time to for each of us to leave Texas for other corners of the United States and the world.

As we had all formed friendships and learned to care about each other, we read our new orders aloud to one another. It seems there were five guys going to Air-Sea Rescue and some going to Electronics schools. Bordeaux was going to Cryptographic school; he was a multi-language guy. There had been three men from the Cookeville area in my training (Billingsley, Taylor, and Smith) but I lost track of them because we were all ordered to different parts of the country after Basic. My orders were to report to the Air Training Command at Chanute AFB near Rantoul, Illinois.

Chanute AFB: AFSC (Air Force Specialty Code) Jet Engine Training School

Lackland AFB (Air Force Base) in Texas was a huge, sprawling complex. Our training barracks were located so far from the runways on the base that to be bussed to the airstrip and loaded onto a C-54 aircraft was a regular thing. I was scared - and I wasn't the only one - as many other recruits were like myself and had never flown *or ever been near* a military aircraft before! I sure was praying in earnest as we sped down the runway and up into the air. I'm not sure how long I prayed as we climbed out over San Antonio, but I know I continued to pray because I was very nervous. The Lord was faithful to keep us safe. It was quite a few hours of flight before arriving at Scott AFB in Belleville, Illinois. When those wheels touched the runway, I was so glad and relieved as we had made it that distance without even an engine flutter. (Perhaps this flying wasn't going to be so bad after all!) We were there but for a few minutes just to pick up and drop off cargo and personnel; then onto the next stop, my job training station.

There was so much going on there at Chanute AFB where I received the next part of my training that I can hardly remember it all. Perhaps things seemed so busy on account of there being a combination of schools occurring all at once such as: special forces, maybe air police, advanced intelligence school, and also the first phase of the air-sea rescue course. The base had several other groups running and yelling, such as paratroopers, each morning.

When I first arrived there at the base in Illinois, I was still trying to figure out the approved Air Force methods and ways. After being issued bedding, I was told to choose a bunk in the assigned barracks so that I might rest my weary head for the evening and, hopefully, the duration of the school. For a reason lost to time, I left my duffle bag and other belongings beside the end bunk. (Later I learned this was a choice position in the military barracks). I had to go to the C.Q. (Charge of Quarters) desk to inquire about some request and, upon my return, I found that a big roughneck had moved my property and taken what I had thought would be my bunk. I asked him to explain himself and his reply was something to the effect, *"You don't own the bunks and lockers in these barracks. If you want this bunk so badly, then you'll have to beat me in a fight to get it back."* Well when fighting words were issued, I accepted the challenge and did attempt to get my space back, but he had a hard and quick left that gave me more trouble than I was expecting so in the end he kept it. Later, I was surprised that he had the gall to ask me for a favor and actually expected that I would jump to! He didn't get his request.

Just like in high school, and academics in general, this job training was not my cup of tea. Therefore, the jet engine course was very hard for me. My grades on tests were barely passing - not only mine, but several other students as well. A lot of us, maybe half, had to go to extra classes at night and we referred to this as the "stupid study". It would not have surprised me if the class instructors had failed a couple of us and sent us into other programs; however, they were obviously evaluating other factors besides just our test grades. To even be placed in that school, the selection board had given us an entrance exam which required average or higher test scores in mechanical and/or electrical aptitude. We spent most of our three months there studying and learning. Besides our courses in aircraft jet engine mechanics, there were several other duties to pull such as two weeks of KP (kitchen police). One of the things that I remember about KP duty was the chow hall. There was a jukebox in our dining area where popular music was played (loudly and constantly) which really got on my nerves after two weeks straight on kitchen duty.

Time seemed to move more slowly nearer to the end of our instruction as all had been away from home (and the majority of the world really) for almost 6 months at this point. We anxiously anticipated the thirty days of leave that would be granted after graduation before going to our permanent duty stations.

Sammy Blain

A significant person I met at Chanute AFB was an airman by the name of Sam Blain from Detroit, Michigan. The military has a pattern of putting people in alphabetical order according to the last name so, because our last names both started with the letter "B", we were paired up for work parties and roll-call formations. This may have been why we met, but I also befriended him because of his personality and his hometown – Detroit. At that time, Sammy and I couldn't have been more opposite. He had a winning, always confident personality, while I was timid and unsure of myself at that point in my life. Another thing that he had going for him was that he had a car and all kinds of friends. And, if you had a car in 1954, that got you more friends than if you had money!

Sammy was streetwise and had been employed for a short time in the auto factories in Detroit, so he was a probably a couple of years older than most of us newly enlisted, 18 to 21-year-old, fresh out of high school or dropouts. He was a good time party person and irresponsible. As our relationship grew, I also found out how dishonest he was, but we remained buddies out of necessity. Because Beverly was now living there, I monitored his activities closely and arranged for him to be my ride to Detroit when we got our 30 days of furlough. Boy, how I had planned for this time!

A few days before we were to leave, Sammy had an accident with the car that broke out the passenger door window and shattered the windshield on that side also. It was the middle of December and "cold outside boys" as Sammy put it while covering the damaged areas with plastic and taping it up. After graduation, on the night before we were to go on the furlough, he went to town and celebrated by getting drunk and staying out all night. So, guess who gets to drive from Rantoul, Illinois to Detroit, Michigan? Sammy said, *"Bowe, you can drive. It's simple; stay on Highway 24, it will take you the whole way. Any trouble, wake me up. I must have some sleep and we'll celebrate more when I get home."*

When Airman Third Class Dorris Ray Bowe saw an opportunity to drive 350 miles, you bet I was filled with excitement! I had very little driving experience (mostly country roads in dad's cattle truck) and the extent of my highway knowledge involved just the 12 miles on Hwy 70N from my small hometown to Cookeville. I was never taught the reading of road signs. The directional signs didn't mean anything to me back then because I was only going one way: either home or to town – Ha! Given how simple my parents were, we children had not been educated or emphasized that the sun rose in the East and set in the West. We vaguely understood from reading books that it did, but that had not been translated to travel. (Silly but true!)

We started North towards Chicago and were to take Highway 24 towards Fort Wayne all the way to Detroit. However, after driving most of the morning in my sleep deprived state, I didn't think to notice directional signs after exiting at a rest stop. I got back on 24 and actually headed west for a good 2-1/2 hours back into Illinois. About that time, Sam woke up and asked me didn't we need some gas about now. His next question was, *"Where are we?! You're going the wrong direction!"* Calmly, I replied, *"Well we're still on 24."* You can imagine that he had a few swear words as we gassed up and headed back onto Hwy 24 - *East* this time. Although I was embarrassed to have made such an easily avoidable mistake, that was a great learning experience. Sammy was upset to say the least, but we did arrive in the big city without further incident. I was excited to get this furlough started as I was going to see loved ones that I had not seen in quite a while.

Over the years, some of the Tennessee Bowe family had moved from the South up to Detroit, Michigan for work. My grandfather, James Bird Bowe, had moved up there for a while in the early 1940's and then my uncles, James Jr. and Alton, migrated North to help with the war effort. My Aunt Dimple was divorced from Henry Gentry and relocated herself, along with her three children, to Detroit. Shortly after arriving there, she met and eventually married a man by the name of Lynn Rupert. In addition to seeing long lost Bowe kin, I was excited to see Beverly Johnson again! We had been faithful to write letters while I was serving in the Air Force, and this was the first time I was seeing Beverly since our parting back in May of 1954. (She had not returned to Tennessee but was attending high school there in Detroit.)

Now it was the middle of December, cold, cloudy, and spitting snow upon my arrival at B. J.'s place. She was so sweet and understanding despite having expected to see me earlier in the day. Sammy drove me to her door and said hello for a moment. Bev's mom was home and we sat in the living room after eating and talked about my uniform and how I had put on some weight, but it wasn't noticeable. The uniform was all that I had to wear at the time because of the allotment going home. Aside from our lifetime memory date, Beverly was dressed up more than I had seen her at school and going to classes. I stayed there at her mom's place for one night, maybe two. I wasn't all that comfortable in this environment because the military barracks had been all I had known for months. Miss Detroit's mother reluctantly left us alone in the house for one day (probably with a stiff warning, given that her mother did not like me from the beginning). We had rubbed each other the wrong way when we happened to meet while Beverly was still in Bixby. I was young and proud, and felt that apology was a sign of weakness, so my response to some comments that her mother had made (which I had interpreted as corrections) was regrettable. Now it seemed as though we couldn't get back on good terms with one another.

After visiting with Bev, I went to Uncle Lynn's and Aunt Dimple Rupert's to visit with them and my cousins: Betty, Jerry, and Kay. These three had gone to school with us in Tennessee. Given that we had been so close, that was quite a reunion. My aunt had been a single mom for several years before meeting Lynn, who was a great guy. Although Lynn was not the biological father to my three cousins, they still called him Dad. Uncle Alton Bowe also lived in Detroit with his wife, Ruth, and children. Ruth didn't visit Bixby often so Alton's family seemed almost like strangers. I believe she was at work, or visiting somewhere, the day I took the bus to their house. My Detroit kin taught me how to get around the city by public transportation. That was very neat; just buy one ticket and get as many transfers as necessary to reach your destination. The ride from her place to Uncle Lynn's was a bit rough, but the Rupert's really helped see me safely to Uncle Al's. It is easier just to tell the bus driver where you were going, then he would pass out a transfer ticket, explain what bus to take, and where to stand so that the bus would arrive on time. On my next visit to Uncle Lynn's during that furlough, the whole family went out to a restaurant for pizza. That was my first introduction to pizza - it didn't sound like something to eat to me. Lynn called the meal a pizza pie. All of these were new experiences for this 20-year-old country boy.

When my furlough time was half finished, it was time to move on so I could spend some time at home. Uncle Lynn insisted that he read my orders and gave me some direction and advice; he was a Marine Corps Veteran and acquainted with military orders. I don't recall who all went with me to the bus station, just that it was time to say those last goodbyes and take the bus ride from Detroit back to Bixby. It would be twelve months before the next furlough. The orders given to me upon finishing the school at Chanute AFB were to report to Mountain Home AFB in Idaho no later than January 10, 1955.

The Visit Home Before Heading to Permanent Duty Station

Now it was December of 1954 and cold outside, but warm and friendly on the bus going back home through Kentucky to Tennessee. All the hometown folk were glad to see me and asked so many questions. Back home, an incident occurred involving my little sister, Carol. She was smart mouthing about something at the dinner table and, reacting out of old habits, I slapped her pretty hard for her disrespectful attitude. I was quite embarrassed, having hoped my time in the military had made me into a changed man - able to have more control and act like the Christian man that I was instead of falling back to the boy who mishandled many situations.

We did celebrate my twenty-first birthday while I was home. Aunt Mamie did something to make the occasion more of a day to remember by having me come to her house where she had made me a special cake. My mother was somewhat jealous of my aunt because Mamie had the time and was able to give us the affection and attention that she was not. We knew our mother loved us but it wasn't as easy for her to show us physical affection.

Orders are orders and my time was up. I now had to leave the Tennessee hills and valleys for the flat sagebrush and tumbleweed country of the West. I often stepped out onto the back porch and wondered if I would ever see the hollows and woods east of Willard Marlow's place again. I didn't know what the future held for me at the next station. Would we be going to war? Will the nation remain at peace? We were in the Cold War with Russia and had nuclear warheads on aircraft flying twenty-four/seven! Dangerous times!! I also wondered what Idaho would be like. The only thing I knew about the state, besides potatoes were grown there, came from a comment made by one of the sergeants at the base in Chanute. He had said, *"There's a woman behind every tree."* Whatever that meant.

The Long Bus Ride to Idaho

So, I boarded the bus at Cookeville, TN, bound for Mountain Home, Idaho, on yet another cold January morning and the first snow I'd seen in my state that winter. We switched busses in Nashville and must have had a straight run or a charter bus continuing on to Memphis. There I saw a sailor and wondered what in the world he was doing so far away from water, only to discover that there was actually a Navy airbase nearby. Once we crossed over the Mississippi River, we passed through Arkansas and Missouri and westward. I must have slept a great deal through Kansas, because the next thing I recall was seeing those high, steep mountains before pulling into the snowy Denver, Colorado bus station. Being in Denver brought back some memories of those old black and white cowboy movies that I grew up watching at the town movie theater or occasionally at school. I didn't get to see much of the city itself, probably because we changed busses there for the next run all the way to Twin Falls. There in Idaho the terrain changed into a lower elevation, now flat and without any trees to be seen for miles; just occasional short bushes. (So now I began to understand the sergeant's reference to the correlation between trees and women – Ha!)

Now I was getting anxious and excited and found a better seat on the bus in order to be able to see further ahead, knowing Mountain Home was somewhere out there and should come into view any time. We were now on such a flat stretch of highway that I could see for miles. For mile after mile, I could see snowcapped mountains far off in distance on one side and nothing but sagebrush and small bushes along the highway. Although I had moved up so I could see the terrain in front of the bus, I fell asleep for a short while and awakened to pretty much the same scenery. Eventually, trees came into view in the distance; this was unexpected since I hadn't seen any since Twin Falls. Yes, it looked like a small forest or an oasis in the desert. Those trees on each side of the street for a couple of miles before we reached the bus station made it seem as if I were back in Indiana. Although I was very excited and my glaring eyes widened over and over as I took in all of the spectacular new terrain, I was simply worn out as well and my butt was numb from sitting,

Upon arriving in the small desert town of Twin Falls after that 2,000-mile bus ride from Tennessee, neither it, nor the bus station was very impressive to this tired soul. It was now January 9, 1955, and I was at the end of my 30-day furlough. It was still miles to where the Air Force base was located in the desert about 12 miles southwest of the town of Mountain Home, Idaho. Being anxious to reach my destination, I asked the clerk at the station counter how to get to there. She said there would be an Air Force bus along any moment now. Suddenly there were others appearing, seemingly from out of nowhere, getting on the bus with me. We rode southwest for 15 minutes before arriving at the base gate. The Air Police Security verified my orders, then I was given some instructions and directions about where to get off the bus for check-in. I signed the pass sheet, was given a map of the base, and a lot of other information such as the hours that the mess hall was open for meals. Boy was I starving, all I had to eat was peanut butter and crackers and that definitely hadn't satisfied me!

There are some things worth noting about the 9th Bomb Wing of the Strategic Air Command (SAC) located at Mountain Home AFB. In addition to the forty-five B-47 Stratojet Bombers allocated to each wing, there were also some KC-97 Aerial Refuelers there with us in the desert. These bombers and tankers were on 24-hour alert in addition to those already in the air loaded with nuclear weapons. The nuclear weapons were to counter the Russian Cold War threat, which was ongoing throughout my assignment there. I thought, "Oh I want to go home, this is too dangerous!" These bombers were ready to go on what could very well be a one-way mission for them - much like the B-25 Doolittle raid over Japan and Tokyo in 1941. The 15th Air Force was stationed in the northwest to be nearest Russia.

Idaho was beautiful in the summertime. there was beauty everywhere you looked. The mountains were a beautiful green with the high blue sky. The very white snowcaps and the bright sun reflection. The winters become gloomy with cold winds.

A view of the snow-capped mountains beyond the hangar

Mt. Home First Baptist

My 1954 Mercury

Mountain Home Air Force Base

The word 'strategic', when first associated with warplanes, had it's beginning in WWII as B-17's flew long-range missions loaded with fuel and bombs to be delivered over Germany. General Curtis LeMay was instrumental in getting the entire Strategic Air Command set up and on alert-ready status to face any threat of attack upon the United States. The 9th Bomb Wing later flew B-29 and B-50's missions over Korea. In the 1960's, they flew missions with B-52's over Vietnam. These Air Force bombers are deployed all over the world to this day. Of course, every place they are stationed, the supporting ground crew and equipment are also deployed to back them up. The General developed the strategy for the best flight formation to go up against enemy fighter planes.

The base had been renovated for the new, six-engine bombers known as the B-47 Stratojet, as well as the KC-97 (an aircraft tanker with four reciprocating engines) which had been modified with a refueling boom specifically for those new bombers. It was such a noisy place what with 45 bombers, 3 tankers, and other aircraft landing and departing seemingly most every minute of the day for the majority of the time I spent there. It seemed odd that these B-47's only had one weapon – a tail gun – but I guess that was because they were so fast that the fighter jets couldn't catch them!

Half of the living spaces had also been renovated inside to become 2-man rooms with a washbasin in them. This Air Force was the place to be! Par for the military of course, new arrivals got to "police the area" by picking up any loose items and garbage around the buildings you are near. They gave us new-to-the-base people some odd jobs to keep us busy until mealtime; or at least until the NCO (Non-Commissioned Officer) for our duty section arrived for orientation classes and job assignments. When I arrived in Mt. Home on Jan. 10, 1955, I was just another A/3C (Airman Third Class) E-2 just out of tech school. (Just another rookie.) It would be another four or five months before promotion to A/2C (Airman Second Class) which is pay grade E-3, and another 18 months before promotion to A/1C – Airman First Class. Confusing, yes?

There were about thirty names on the orders which included my friend Sam Blain. We had even been assigned to the same barracks. Sammy and I got into a lengthy confrontation about our recent Detroit visit and the funds he owed me. Even though he was now driving a different car, he was still the same old guy; ready to go partying and borrowing money from any unsuspecting person to get the process started. I cannot say when exactly, but while I was still new to Mountain Home AFB, the Lord put it on my heart to rededicate my faith life back to Him. Additionally, He put some new goals into my life: I must get to know Him better, trust Him more, a new beginning to walk daily with the Lord as I searched to know His will for my life as an airman. I thought it wise to meet the chaplain and his staff if he had one. This was a good place to start over new. After having grown up in a large family, an almost 3-day bus ride had left me feeling pretty lonely. We may know Jesus is there but we also want a person near.

First Work for the Master: The Chapel Annex at Mt. Home

Well, Chaplain Dalton did have several helpers for the many children's Sunday school classes already in progress. I knew from Saint Paul's letters, that as a believer, we are supposed to work for our Lord, so I began trying to teach, but mostly just supervised and directed the children by trying to make them welcome and comfortable in the church environment. Believers always find each other and there is no mystery there. God is an all-knowing God, and He knows our needs before we ask. The men I met on base went to church in the town of Mt. Home. The church was a very friendly place with special things and events for members. To me, those words were like music to my ears: friendly, welcome, with special events and programs for all ages. Bill Harris, a married airman living on the base, promised to pick me up and give me that ride in and back each Sunday that I cared to go.

The Dirty Shop with Heavy Tires

Extremely strong airmen were used at the Fuel Maintenance Shop located at the back of Hangar One. Here, the various tires and wheels from the B-47 and the KC-97's were regularly serviced and repaired; two or three planes at once inside and at various locations on the ramp.

The Electric Shop was constantly bogged down with gremlins to the point where they could never spare any manpower. Teams of two airmen would be tasked with tracing the wires with the aid of a diagram. Airman Third Class George Macintyre became experienced in finding and removing these small wires that extended inside the full length of a bomber's wing. He and I worked third shift tracing wires on one aircraft for 2 weeks or longer . Sheet-metal men removed wing panels for us to access the wiring harness and its attachment to the wing ribs and spurs. The electric and instrument testing equipment may have shared some spaces, but the large spools of so many different sizes and shapes of wire that seemed to go on for days required a lot of room.

The Engine Shop had the largest pool of airmen; really more than the Maintenance Shop could use and efficiently train all at one time. The NCO knew that airmen were going to have to be cross-trained as other airmen are transferred. Next is the Parachute Shop which packed the personal chutes for the aircrew. Both personal and drag chutes had to be stretched almost the length of the hanger for inspection and repair. 'Drag chutes' are parachutes used to help slow the aircraft to stop it from going off the end of the runway and they also help save the wheel brakes. Every time a B-47 landed, it deployed a brake chute and a truck and crew would go and retrieve it for re-packing. Technology did not prevent chutes from failing to deploy on rare occasions.

Then there was the Machine Shop - airmen deployed to this shop generally remained there. Unless a broken bolt needed to be extracted, we never saw any of them. As a matter of fact, in 2 years, this only happened once. At the west end hangar of the flight line was the Hydraulic Shop which was responsible for all the hydraulic cylinders which controlled the actuator. The actuator controlled the aircraft's vertical and horizontal stabilization, flaps and landing gear. This shop was quite busy because they also handled the landing gear since there were some huge cylinders used there.

The Ground Power Section

After some utility assignments, some of the other recently schooled Jet Mechs and I were reassigned to mech training duty in other shops. Nervous and excited, we viewed the aircraft being loaded with the first atomic bombs right there in the 9th Wing of the Strategic Air Command at Mt. Home AFB. The first place that we were sent was the Ground Power Shop. This was just one of the many repair and supply shops in what we called 'the next-door hangar' that was about a quarter mile from the barracks. The ground power units were starter units for the jet engines we had been introduced to while at the school in Chanute, so I had some training on them.

Although I assumed I would go to the Jet Engine Shop first, I was somewhat disappointed when the top NCO loaned me out to a different position in the Ground Power Shop. In the shop interview with Maintenance NCO Master Sergeant Harmon, he had a folder detailing the short history of my time in service (basic training, school grades, etc.) and wanted to discuss the importance of education, which was high on the Air Force's priorities. Harman mentioned the fact that I did not have a high school diploma and that there was some work I needed to accomplish before the next promotion cycle if I were considering making the Air Force a career. He explained the training program briefly and said, *"We need you to spend some time in the ground power section, which is part of the job and career process."* He gave me a paper with the NCO's name and pointed East from where we were in the elevated office inside the main base hangar. He told me to report over there to Sergeant Giles in the morning. I started to salute, then awkwardly put my arm back to my side because he was not an officer, and said, *"Yes sir, thank you."*

So began my career in the maintenance side, but like any apprentice I started on the ground level or worst part of the job first. There was not a great deal of work for this newcomer A/3C. This was an OJT (On the Job Training) field so most of the men that worked here were considered "low-skilled" and had simply been taught their jobs. (While being extremely important in that these tasks get done, my opinion based on my brothers' experiences is that these OJT jobs are not quick or fast promotional career jobs.) Along with my new assignment, when my commanding officer found out that I had enlisted without a high school diploma, I was encouraged to participate in a new GED program along with my regular workload.

Airman Martinez and the Power Generators of the B-47 Aircraft

This is where I met Martinez, a genuinely nice young kid barely old enough to be in the military. This man laughed and carried on so much we hardly knew when he was serious. As I have said before, there was not much to do with these units until they were needed by an aircraft doing their pre-flight checklist before the start of a mission. There was an office in the shop where the officer in charge could update the workload sheets and sign the maintenance forms. When the NCO was out of the shop, we were required to answer the phone. Martinez and two other airmen were busy with their non-duty activities when I found a typewriter and started practicing my typing skill (actually I was doing some paperwork for the NCO). Because of the typing, Martinez gave me a new name, "Bowe the tit-less WAF" (Women Air Force). We laughed for quite a while about that. Laughter became commonplace while working with him because he could make a joke out of most any mistakes that one made.

The Ground Power Units were associated with the Engine Shop more than I had imagined. The Jet Engine Shop had a test stand mounted into a cement bunker about a half mile away from the shop. We had a four-man crew that would switch out an engine. First we would have to take off the nacelle covering the engine, and then just switch out one engine for another like an auto mechanic would switch out a transmission.

When the new engine was attached, we would run power cords and lines from the power unit for diagnostics to get the engine started. That operation would take more manpower than one would think; four men using a tug or a pickup truck to transport the engine out there, then connecting all those hoses and cables accurately and quickly. If you have ever watched the mechanic put a transmission or a motor in a car, imagine doing so with an airplane engine as big as a station wagon.

As soon as all the lines and hoses are secured, now is the time to call for the power unit which is the size of a VW Bug. The crew for this consisted of three men: one operator, one as fire guard, and the other to help position the unit and hook the power cables. In the extreme cases of things going horribly wrong and lives are lost, one well-trained airman could operate the power unit with help from the engine crew. In the civilian world, a ground crewman would bring the unit out in the back of a pickup truck and hook everything up by himself; but we are apprentices learning the job. I stayed 90 days there before going to the Jet shop.

James Mason

A/1C James Mason became our group leader. He was an older person, one of the married airmen in the Jet Shop. He demonstrated some strong mechanical skills. Instead of on base in the barracks like most of us, James lived in the town with his wife and two children. He would invite a few of us that he trusted to his house for some home cooking. Mrs. Mason had a warm and friendly personality. An educated and well-read woman, we enjoyed her conversation as much as her meals. She had a knack for quick comments sprinkled with humor. Glenda was witty, smart, and fun to be around - always full of Midwestern slang and jokes. To see her and be around her, we could very easily understand why James had chosen her for his wife. After a few visits to his place and her home-cooked meals, we came to love their fine home and especially enjoyed hearing her make wisecracks about the Air Force and the military in general. James and Glenda were a wonderful couple, full of fun and life.

The four men that carpooled to the worship services at First Baptist Mountain Home along with me were all unique in their own way. First one was a tall slender airman who came to church in my last year there. Second was Elmer James, another from our shop. He was a genuinely nice Christian man from Virginia who had a real southern drawl. He was one who always talked about guns or "Hog Legs" as he called them. For some reason, Elmer would only talk to a girl when another man was present; mostly for appearance's sake. Otherwise, he did not want anything to do with the female gender. The third airman was a Baptist Minister's son who wanted all the girls to like him but went about it in an offensive way. The fourth one who regularly rode with us did eventually get engaged to one of the local girls there.

Life on the base was blah, dull and noisy with those bombers constantly taking off and landing. Even if not for specific missions, there were landings and takeoffs for practice. We as aircrew refer to that practice and procedure as 'Shooting Transitions'. The new pilots had to be trained and obtain their in-flight hours. The pilots had to make a minimum number of landings and take offs, have a minimum number of flight hours (per month or quarter) per year to stay qualified as pilots. The 'shooting transition' was the repeated practice of taking off from the ground, turning around, and then landing.

Simple daily maintenance of clothes, washing, sewing buttons and stripes on both the work and dress uniforms, etc. had to be done. Maintaining the military uniform standards were most important, but getting a regular haircut and polishing shoes seems dull compared to the glamour and excitement of traveling to another base or country.

Free Time at Mt. Home

God provided such beautiful sights and wonders everywhere, including the Idaho Desert. I had several occasions to ride and drive to the cities of Boise, Nampa, and Caldwell. A friend drove me up a mountain that overlooked Boise; it was a beautiful city. Seeing the city prompted me to wonder about the people and its history. A bit of research told me that in the 1860's, Boise was a gold mining town. They mined the mountains surrounding the city set in that beautiful valley. The ranchers and miners got their supplies in town: Their working tools and equipment, food such as fruits and vegetables, and the famous potatoes. There were beautiful places to visit and to live, providing that there was work to be had. I never recalled what the main industry was in Boise during the 1950's. There was a National Guard Battalion; the Air National Guard unit had recently lost a couple of pilots. Their new jet fighter was about to crash so both pilots bailed out, but one pilot's parachute only partially opened as the plane was going down. The Boise Time newspaper was delivered to the base, so we read several stories of the happenings in and around that town.

There were some civilians from Boise that worked in the Jet Shop with us. One was concerned about his daughter and the men she was attracted to. Let us call him "Walter" because he reminded me of the actor, Walter Brannon, of western movie fame. Anyway, Walter invited me to go home with him to introduce me meet his daughter. (Young ladies very seldom like the men that their fathers would prefer them to be courted by.) Walter's wife prepared a nice meal for the whole family, including me. Although Mrs. Brannon seemed to like me as much as her husband did, their daughter had already chosen a different course of action and stayed out late that very night.

Simply because the chow hall never had a home-cooked meal, spending the night under Walter and his wife's roof was such a treat for me - although awkward because the purpose for my being there was not there herself. As Walter and I made small talk on the way back from Boise to our place of work the next morning, it shortly became evident that the topic of his daughter was not up for discussion at this moment in time.

Because of distance and availability, the methods for approaching and meeting and/or dating the girls of the time were different at Mt. Home than they were during my training in Illinois. I would have to say Sammy Blain's methods and thoughts were the same no matter where he found himself: "Go to a bar (or several bars) in the nearby town, find the loose girls "looking for a good time", and drink as long as the money held out.

Airman Thomas was from another unit on our base whom I met and befriended during my time at Mountain Home AFB. He and I used the more modern and modest approach when it came to seeking female companionship. As our friendship grew and we had more conversations, things would come up about the fun activities that we both loved to do. We found out that there was a roller-skating rink in a small town nearby, and we thought perhaps there would be many young people since the Northwest

College of the Nazarenes was located there. Thomas had an MG convertible and he loved to zip that small car around in small places. We had so much fun driving to the skating rink in Nampa which stayed open late. We would go Friday nights and stay until it closed, then get home in the wee hours of the morning. We met and skated with a lot of girls but didn't find any keepers. In my observations, I have come to believe that there are ladies everywhere that want to "spoil the fun" by keeping a guy from the young lady that he would prefer to pursue - the difference between the one you like and one that likes you. This one girl would try to flirt with Thomas the entire time we were there at the roller dome. He would try to shake her off and politely discourage her from acting on her attraction to him. (Neither one of us considered her a beauty in appearance according to our standards at that time.) After a few attempts to skate despite this girl's crush on Thomas, we thought that it might be better to go exploring for new places to meet young, college-aged adults. Besides someone to go skating with, we also enjoyed miniature golf and he was glad to have someone to share the fuel expense and laugh at his silly jokes.

Nick Greer was one of the airmen from the shared barracks in which we lived. From Florida, he was a proud, boastful individual - especially concerning his accomplishments in sports and with all of the women he claimed he left crying at the airport when he left home. Apparently, he had been a star quarterback on his high school football team a couple of years before. He could not stop talking about his participation in several sports and very openly expressed his thoughts about girls and dating. Greer seemed to feel intimidated by me and always wanted to compete, especially in the way of conversations about our pasts on the football field. He even challenged me to a foot race so he could prove who was the better man. Nick won by quite a stretch, which I knew he would, but I just wanted to make him run on account of his constant boasting.

Despite our confrontational relationship, my desire for adventure was strong enough to make me accept a ride to Boise with him. We drove around the city, slowing down as we passed the drive-thru and car hop diners long enough to see what type of crowd gathered there. Those diners were very popular in the 1950's and 60's. Greer was compelled to stop at this one diner when a very beautiful girl, dressed in some kind of feminine fashion overalls, caught his eye. What a knockout she was, so Nick just had to talk to her! I thought she would ignore him and just to fall back into conversation with the other boys she had been cutting up with, but Greer finally talked her into going for a ride with him in his car.

They were gone for quite a while, and when they came back were all affectionate and seemed very happy with one another. Nick let her out and I got back into the car. We headed across town to visit another diner in hopes of picking up one of the ladies getting off work. Little did we know that the girl in overalls was diner hopping too because she also showed up there. I was not in earshot of their conversation, so I don't know what words were exchanged, but the girl started crying big tears. I was embarrassed to be with him at that moment. That was my first and only experience going out on the town with Nick Greer.

The First TDY: Off to England (45 days)

The military has a special way of life that all personnel learn to endure. It is called "Hurry up and wait." The experiences of my first year in uniform gave me the opportunity to witness how this works while on three different Air Force bases. There was constant movement of personnel, behind which ran a well-oiled rumor mill that buzzed with possibilities of exotic destinations and dreams of receiving orders to a Pacific Island paradise, etc. I watched what looked like mass confusion as people were constantly being shipped off to different bases with different orders. Did someone know what was going on?

This was the beginning of my second year in the Air Force. General Curtis Lemay had designed a 'System of Readiness' - meaning that we were to be ready for deployment at all times. Within 24 hours of the Strategic Air Command (SAC) giving the signal, we could be on our way to any one of many forward SAC bases in Asia, Europe, Africa, Canada or Alaska. We were well trained and ready at a moment's notice.

Finally, the 9th Bomb Wing to which I was attached experienced the difference between truth and rumor. We received orders from SAC Headquarters in Omaha, Nebraska: *"Put the war plan into action."* That meant that we were to load aircraft with the special 'Flyaway Kit', fly to a forward base, and conduct mission strikes from there. Our orders were to proceed to the Royal Air Force (RAF) Base in Fairford, England, to join an RB-47 bomber group. So, like birds setting out on a seasonal migration, all forty-five of the 9th Bomb Wing's B-47 bombers departed Mountain Home AFB. Boy was it noisy there on the base for the next three to five hours! The air was so thick with smoke and fumes, it looked like a dust storm. It took a few days before we were ready to follow the bombers, but at last it was time for the maintenance squadrons to assemble their men and equipment. We proceeded to load up onto the big air transports.

In the weeks and months leading up to this TDY, I was not that excited about traveling outside the U.S. Since I was still rather uncomfortable about flying, I felt a need to visit the base chapel. And it was there, for the first time in my young life, I found an opportunity to exercise my faith and begin fulfilling the "Great Commission." The words of Jesus, as written in the gospel of Matthew 27:19-20 and Mark 16:15-18, command us to go into the world and teach. Soon after taking that command seriously, I found that I was being used to help one of the airman's wives teach the children at the chapel annex. This sweet lady really knew her Bible. She loved the Lord and God's love for the children came shining through her tireless work with them. The children under her care were mostly those of base personnel, although some of them must have been from non-military families. One thing I have noticed is that people of faith find places for their children to be loved and taught of the Lord. They will travel miles for their children to have the benefit of a special Godly teacher and mentor.

Another young man who I met about that time was also growing in the Lord. We worked together helping with the Sunday school effort at the base chapel. Somehow, he sensed that I'd be happier attending the Baptist Church in the town of Mountain Home rather than the one there on base. I loved the worship service in that new civilian setting! We sang familiar hymns and enjoyed fellowship with a greater variety of people. Some of the older airmen who had transportation attended there too, so all was well. The twelve-mile drive to and from town wasn't that much of a thrill, but even that was more pleasant than teaching every week at the base. [Perhaps I ran away from what God was asking me to do there? This may be a good example of how people are swept away into a more social gathering rather than doing God's work and being the priests they are called to be, but I digress.] Our Bomber Wing had flown the coop and all of us in the maintenance crew needed to follow as an important support team.

Orders are orders and we had to leave our most enjoyable church culture behind for about two months. We had many questions: What will old England be like? How will it be different than what we are used to? What will it all look like? After spending the past several months in the desert, will we miss the tumbleweed and sagebrush? Will we miss all these wide-open spaces that we have grown to love in this part of Idaho? Some of the airmen speculated that England would be more like Boise and the Nampa Valley. Well, we just had to wait and see.

But first a little problem - probably par for the course in the United States Air Force. In preparing to go, we found a J-47 engine in need of repair that was left behind in the shop. It should have been loaded and taken with the other supplies, but maybe it was left because of the minor work needed. Our responsibility was to get down to business and do the repair even if it took all night. With all hands on deck, and even supervisors getting their hands dirty, it didn't take long to get that engine in ship-shape condition ready to load the next morning.

It had now been two days since the first group of maintenance crews had taken off for this new deployment and we hadn't heard a word from them. Even the ever-present rumor mill seemed to have quit cranking. In the chow hall at mealtime, I brought up the subject regarding the lack of information about our buddies enroute. My fellow airmen must have sensed my growing apprehension about the long flight ahead of us as I nervously questioned previous crews about their experience flying to forward bases. Like all young men, they wanted to tease and have a good laugh at my expense, so they started spinning a yarn about how two of our planes had already ditched at sea! On seeing my worried facial expression, they all erupted into laughs and guffaws before they said, *"We are only kidding, Bowe."*

Finally, it was time for our group to board the Navy R5D passenger transport. Navy R5D was the designation for the aircraft that the U.S. Air Force called the C-54 Skymaster. That type of transport is what the Air Force used along with several other types of transports to supply the Germans and East Berliners during the Berlin blockade of 1946-47. We thought it odd that the seats in the Navy R5D were facing backward instead of the normal seating arrangement that we were accustomed to. The Navy had ditched at sea often enough to know that this rear-facing seating arrangement enhanced the survival rate in crisis situations; after all, passengers didn't need to see where they were going with a Navy crew flying over the ocean. And so it was that we left the runway at Mountain Home AFB with a smoother takeoff than the last one we experienced in that old C-54 at Chanute AFB, Illinois.

After flying for several hours, we landed at RCAF Station Shearwater in Nova Scotia. The base seemed like a strange place – as it was without paved streets or sidewalks. It was late in the spring of 1955 and very, very muddy as we walked to the chow hall and then on to the store known as a Base Exchange (BX). For the men permanently stationed there, walking in the mud seemed normal. After all, they had just come through a long winter with lots of snow and slush on the ground even into June.

After we made the Nova Scotia pit stop for refueling, we were reloaded back onto the Navy transport and couldn't help but notice that a survival vest was now loosely attached to the back of each seat. After buckling in, the Navy stewardess briefed us all on the ditching procedures by demonstrating the proper use of the survival vest. Down the runway we went and up into the leaden skies for what we thought would be a very lengthy flight across the Atlantic Ocean. The flight wasn't that long, however, since we only flew halfway across the Atlantic - landing on the small Azores Island of Terceira.

The Azores, (officially The Autonomous Region of the Azores) is a self-governing region of Portugal; an archipelago composed of nine volcanic islands in the North Atlantic Ocean about 850 miles west of continental Portugal, 936 miles northwest of Morocco, and 1,196 miles southeast of Newfoundland, Canada. We landed in the dark at Lajes Field on Terceira Island which was the home of the 65th AF Wing. This tiny island played a huge role during the 2nd World War as being the place where President Roosevelt and Prime Minister Churchill met and discussed how they might acquire a base in the Azores. Portugal wanted to stay neutral, but as the war developed there were fears that the Germans were going to take over the islands. The Portuguese government constructed some buildings and extended the runways knowing they couldn't defend the islands with the small forces stationed there. So, Portugal joined the allies and sent troops and equipment to England. These islands were a vital strategic crossroads in the Atlantic during World War II - used effectively as an important base of operations by the Royal Navy, the US Army, and the US Air Force. In fact, numerous missions using the B-24 Liberator aircraft were flown from these islands.

Because it was dark, I wasn't able to see how beautiful this place was when we deplaned and were led to the mess hall. Interestingly, our meal was served from bowls instead of the usual plates and cups. There in the mess hall, native Portuguese did the menial jobs such as the serving and cleaning. They went quietly about their work; we didn't hear them speaking. As day broke and the sun came up, we could see that the island we had landed on appeared to be a huge rock, just large enough for a few buildings and the air base. We stayed the next night for a scheduled time of rest and recovery because there are usually some cases of air sickness as a result of that 12-hour long flight from Mountain Home.

RAF Base, England

The next leg of that long trip brought us to our temporary duty station: the Royal Air Force (RAF) Base in Fairford, England. We arrived stiff and tired. Who knew that sitting in a plane for so long would take that much out of a person? But that didn't matter to the sergeant in charge as he loudly barked, *"Get over to the hangar, grab your gear and your toolboxes, and when you're finished with that each of you get a broom and cleaning equipment."* Our new work area had to be cleaned just like the deck of a ship. The NCO in charge yelled out, *"There is an engine and all its parts that haven't made it in the hangar yet. Clean this place up and bring all the equipment in, then get to work on it."*

The next morning all of us new arrivals were herded into several meetings about our new duty station. They lectured us about all sorts of things including the venereal diseases that were rampant if we were not careful when fooling around with easily available English women. After that we were required to check the message board for our duty station and find out what our particular job was. One look at that board and I couldn't believe it! Crushing disappointments truly are a regular part of life - especially military life. They surface when you least expect them, but I took consolation in thinking about the Biblical story of the life of Job. He lost his entire family, which must have been heart wrenching for even a Godly man. Job knew God and remarked later, *"The Lord gives, and the Lord takes away. Blessed be the name of the Lord!!"* Job 1:21b

My disappointment stemmed from the fact that I was a trained jet engine mechanic but, as a new arrival here in England, I was viewed as an extra. I had the distinct feeling that I wasn't really needed. That impression was confirmed loud and clear when management assigned me to the Ground Power Shop, then quickly moved me to the Engine Shop, and then to Base Security. Of all places to put a jet engine mechanic! I was given an M1 Carbine, a clip of ammo, and the orders, *"Do not to fire unless fired upon!"* Adding insult to injury, I was placed out in the boonies on 3rd shift to guard B-47 bombers from saboteurs - one of the airplanes I was supposed to be repairing!

They were quick to brief us on proper care and use of an M-1 carbine rifle as well as all the other procedures that were expected of us. We then had to fall into a formation made up of other "excess" troops and marched to a briefing where we were directed to view the base map as they described what was expected of us as a security detail. Some details were heavily emphasized, such as when and when not to fire our weapon. This was a base that had withstood World War II but was now being threatened by the Russians. This was the time of the Cold War just after England had joined the North Atlantic Treaty Organization (NATO). The base was laid out in such a way that an enemy aircraft could do little damage by flying a straight-line strafing or bombing run. The circular or diamond-shaped parking pads, separated by considerable distances and odd angles, would force an attacking enemy fighter to make multiple passes.

The Strategic Air Command (SAC)

The Strategic Air Command was always poised and ready to strike anywhere, anytime. General Curtis LeMay was responsible for having the many groups in position to do more damage to the enemy than they could do to us. Korea and Vietnam were limited wars because both the U.S. and Russia feared destruction. There were many alerts and false alarms. The Fifteenth Air Force, of which the Ninth Bomb Wing was a part, was always either preparing to go to a forward base or they were there carrying out simulated strike missions.

The heightened tension of the security detail made it quite stressful. In fact, it was no fun at all to walk patrol around B-47 bombers parked on pads spaced two football fields apart. We had been warned that a saboteur had breached the fence on another RAF base and had blown up one of the planes there. Not allowed to talk or even wave at a fellow sentry, we found it strange that the darkness only lasted 3 hours; and even then it was more like dusk throughout the night. The sergeant or corporal of the guard came driving around in a jeep every couple of hours asking if we had seen or heard anything and reminding us to keep watching the fence. Watch we did. We saw lots of movement; the local bunnies seemed to take delight in startling us. It was somewhat disconcerting being amid the hares out in those British boondocks in the dark.

SAC Wing Organization

The Wing was SAC's basic operational unit. It reflected a major change from the previous organizational structure. During World War II, the basic combat unit was the group and groups were assigned to bases. Often the base commander was a regular army officer with no flying experience. This often led to many problems. On September 16, 1947, the United States Air Force was established as a separate and equal element of the United States armed forces. One of the first things it did was implement the Hobson Plan for reorganization. Often called the "Base Plan," the Wing Commander was also the Base Commander.

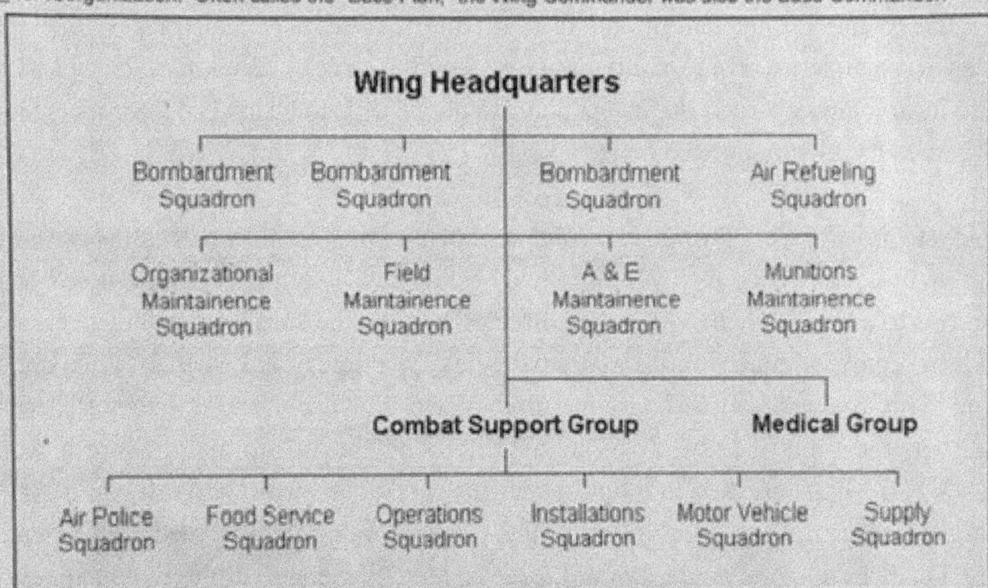

The above chart shows the organization of a typical SAC wing of the 1950s. A wing was generally assigned four tactical squadrons - three bombardment and one air refueling. The primary responsibility of the Wing Commander was to provide combat leadership. The tactical squadrons were supported by four maintenance squadrons.

"Hell-o England, Goodbye Heart"

Miss Detroit and I had been exchanging letters for over a year, both in high school as well as after I enlisted. Even though one or two other women had tempted me to write them, in my heart nothing felt as sure and comfortable as my feelings for her. I was certain of it - Beverly was the only woman for me. I was always so happy to receive those love letters that I kept them for years. I kept them even after receiving a final heartbreaking letter. I must have read that last letter a hundred or more times, but unfortunately it always said the same thing, "Goodbye." All of her previous letters had always started out with "Honey", "Sweetheart", "Darling", or something to that effect. This one started with "Dorris" and stated, *"Do not write, call, or even try to look me up – Ever. Goodbye."* Unable to believe what I had just read, my heart and my life seemed to have suddenly stopped. I have never had a cardiac arrest but this must be the closest thing to it! That letter was a shock and a thousand thoughts ran through my mind. "What did I ever do to make this letter sound so cold?"

With the change of tone, so very different from all her previous letters, it made me wonder if she had actually written this one. I could easily imagine her mother writing it. She and I had had some differing opinions when we first met and it's hard to win the girl when her mother is an opponent. What a sad and disappointing day that was; made even worse by the fact that it happened while I was posted overseas. That awful letter stayed with me day and night. I couldn't sleep, couldn't eat; my peace was stolen. Worst of all, I didn't have a believing Christian friend that I trusted to share and pray with. Life in general became lonesome. I just wanted to go home and cry myself to sleep. Knowing how earnestly I had hoped that my investment in the relationship with Miss Detroit would result in marriage, I cried a lot. The timing and place could not have been worse. Suicide? Yes, that was considered briefly, but God is faithful and has a plan when all seems lost. In the end, I found refuge in the Lord. Thank you, Jesus! (The anchor holds, on Christ the solid rock I stand, all others are sinking sand.)

People will fail you, but I had my faith in God and the Bible to comfort me and tell me I hadn't done anything to deserve this. There was a reason for this heartache. There is always a reason. The problem is that the reasons behind any difficult situation remain unknown until the proper time when all things are revealed. Did it matter? My mind raced. God had blessed me by giving me many friends since I first met Beverly. The letters and notes Beverly and I shared over those two years was a growing experience that I will never regret because God would eventually use it to shape me.

As I prayed about it more, God began to show me that he had a great lesson on the importance of patience He wanted me to learn. We had expressed the warmest and deepest feelings for one another on paper. At that time of my life, Beverly was the first woman that I could express those thoughts and feelings to. Love is wonderful even if it is fleeting, lasting only for a short time. When you really, truly love someone and they betray or disappoint you, you may briefly feel angry, but then you wish them well and still want the best for them. Now I had to put this sad development behind me and move on.

The Brits

Every new deployment presents opportunities to see new places and experience new things, such was the case while stationed under the Union Jack. Old England was a nice place to visit. I was happy to be there even when we could have been stationed at any one of the many SAC bases in Spain, Turkey, Saudi Arabia, or perhaps even Alaska. We were amused by the different ways these English went about doing their daily work. For instance, some of the menial day laborers on the base would ride a bicycle from home to the base dressed in a suit and tie. They would ride the bicycle to the tool storage shack where they would take off the tie, roll up their sleeves, choose between either a pick or shovel and then begin to dig either a trench or a hole depending on their task. This was something that I witnessed more than once on a fairly regular basis.

British women also rode bicycles, their dresses or skirts flying in the breeze provided plenty of skin to admire as they pedaled past with their bare legs pumping. Were we soldiers supposed to look the other way? We didn't, which may be one reason why the Brits put out a well-known and well-earned statement about American servicemen during WWII: "The Americans are oversexed and over here."

My heart was often in turmoil as God prepared me in ways that I wouldn't understand until looking back upon things later. With that hindsight came insights on various coinciding events, things that were later revealed to me. One of those provided some levity at the expense of a sergeant who chose to be a mean jerk to the RB-47 group. It is quite a story. Like many tales, this one starts in a British pub.

Two class E-7 Tech Sergeants, both assumed to be family men as well as 15-year enlisted men, were drinking at the Non-Commissioned Officers (NCO) club. These two increasingly intoxicated sergeants were unable to handle their drink. Eventually a challenge was issued, although no one knew from which foul mouth it originated. Nevertheless, the challenge was out there between the two drunks and involved a joyride in a half-track. Now a half-track is a vehicle whose front half is that of a pick-up truck but, instead of the back tires below the bed of the truck, it has tank treads designed to handle extreme weight (much more than a regular truck could handle). For many years, these were the Army's all-terrain vehicle. At this RAF base, they were used specifically for dealing with snow and ice on and around the ramps and flight lines. As of 2019, there were still about 600 of them currently being used by the military.

The idea was to "steal/borrow" a half-track from motor pool and joyride it around the base during the darkest part of the night when they thought everyone would be asleep and no one would be around to catch them. But being skunk drunk, they didn't wait to pull off their stunt. There were plenty of onlookers and everyone involved was having a raucous good time witnessing this half-track being driven by the drunk sergeants. They drove everywhere that normal automobiles were not allowed to go - even where they knew they should NOT be driving it, like around the flight line and across the runway several times.

And like most situations in a competition, one is inspired to outdo the other; so, when the initial task had been accomplished, the mental haze of intoxication had one of the two sergeants involved decide to "step it up a notch". His drunken imagination must have gotten the better of him. All that anyone could tell in the dim black of night was that one of the two inebriates had managed to crawl into the bed of the vehicle over the tank treads and was standing up in it pantomiming a charge of the old 7th Cavalry. (The 7th Cavalry was no longer around as it had been folded into the 1st Cavalry division in the 1920's.) At first, the stance and antics by this sergeant just added to the fun that everyone was having. Then, the driver made several sharp turns as well as going 'all-terrain' in the dark which resulted in the standing sergeant tumbling out of the half-track's bed. As the old saying goes, "It's all fun and games until someone gets hurt", and this sergeant was definitely hurt when he fell onto the pavement. His injuries necessitated taking him to the base hospital for treatment.

The next morning proved to be the beginning of the end for these two hungover sergeants. Because of the injury, what was supposed to be a secret challenge and prank wound up becoming a full-blown military investigation to determine whether or not these two men could've pulled this stunt off alone. At a minimum, it was asserted that they must have had help "borrowing" the half-track from the motor pool. Since our maintenance squad was only there for temporary duty assignments, we did not have the clearance to be privy to the scope of or progress in the investigation. Along with much of the rest of the base, we were dependent upon the rumor mill for information; however, we did witness the punishment meted out as a result. As their public punishment, each of them was ordered to dig a hole 6 feet wide by 6 feet long and 6 feet deep next to the chow hall. Witnessing these two sweating, dirty, hole-digging sergeants performing their punishment in public gave all who were waiting in line for their next meal an extended opportunity to think before they act. While that was probably the primary reason for their public consequences, having to listen to the snickering of their subordinates watching them as they peered out of their deep holes was definitely a thoroughly humiliating secondary lesson stemming from getting caught being stupid.

These are supposedly us young men's leaders? Strange times and events; even our Chaplain from the 9th Bomb Group was seen trying to stumble to his quarters. Wow! This was not America and we had not been devastated and almost defeated by WWII like England was before America did the lend-lease program.

While the alcohol-induced levity from this half-track episode gave a brief but welcome respite to my heart, which was still reeling from the personal tragedy of my failed love life, the reality of our 9th Bomb Squadron being stationed in England remained ever present. The Ninth was over here to defend England and help with the Cold War efforts to contain the USSR by flying protective and intelligence gathering missions. I clearly remember one bullet and shrapnel riddled RB-47E returning from a reconnaissance mission. The plane was a Boeing B-47E Stratojet (the 'R' in front of the B-47 stands for Reconnaissance). Upon landing, this crippled plane was quickly escorted by the Air Police and other security vehicles to a special secured hangar. While in there, security around it was at the highest level. Later we learned that the plane had been flying reconnaissance near Romanian airspace which was controlled by the USSR at that time. The mission of this flight was to take pictures of Soviet activity in and around the Romanian border. Apparently, the Russians didn't appreciate our boys knowing what they were doing - hence the bullet holes in the plane.

The RB-47 was designed to replace the RB-29 and RB-50 Superfortress aircraft which had been serving in long-range photo reconnaissance roles, but both of which were rapidly approaching obsolescence. The RB-47 carried out many ferret missions around the periphery of Soviet territory, and sometimes on penetrating flights inside, to map planned routes for the B-52's should combat missions over the Soviet Union ever become necessary. Beginning in 1953, the first production models of the reconnaissance B-47E were delivered to the USAF with the last of the 255 of them delivered two years later in August 1955.

How can one deployment contain such a variety of fun and stress-filled days all shoved together? Each day was an adventure that could lead to a different outcome. With all its ups and downs, this was truly an emotional roller-coaster time for this young airman on the Christian journey.

In that stressful environment, I'll never forget how we were allowed to let off steam and relax. Many of us were given passes to go off base and explore, finding what was new and interesting in this country so very foreign to us. Some of the airmen who could afford to travel the 90 miles to London took long trips there to unwind. All of us had heard all kinds of stories about what goes on at the renowned Piccadilly Circus - the red-light district where prostitutes sold their services. Since I never went there, I can only imagine how those girls dressed in order to seduce servicemen.

Although my heart was still aching from my "Dear John" letter from Beverly, I knew that I had to get back on the horse in terms of relationships, so I attempted to befriend some British women. Boy, was I wrong in expecting them to have the same hard-to-win attitudes that I had experienced in America! Given that England was a war front throughout the Second World War, most of the allied side was "baby crazy" after the victory against the Germans. More than 50,000 pilots had been lost, as well as the men who had died in the ground fighting and navy sea battles. As a result, for a huge number of women in the United Kingdom, eligible and desirable men were few and far between. For these reasons, British women were much more aggressive about meeting or picking up men than I had experienced in America.

One brief experience with the women of the U.K. came when I went to the guard shack at the main gate of the Fairford RAF Base. My purpose in going there was to vouch for the identity of one of the airmen in our 9th Bomb Group. Until my services were needed, I interacted with and met whoever crossed my path. As I mingled, some of the women around the shack asked if I knew one airman or another. I could tell that several of them were just not interested in me, which didn't bother me one bit. They explained that they were there waiting on their lovers to show up so they could get on with the rest of their planned activities. One woman seemed alone so I moved in expecting to have to engage her interest. Much to my surprise, she turned out to be one of those extremely forward female Brits and quickly began trying to chat me up. She told me that she was looking for a lover, specifically a relationship. She thought that she was flirting with me as she kept pointing out differences between our two cultures.

You see, when on leave from the base, we airmen dressed more casually. One way we often did this was by leaving a few of our top shirt buttons undone, showing a bit of white t-shirt underneath. This forward young lady trying to butter me up noticed and coyly inquired about “showing off your underwear” - the implication being of course, “If you freely show off the 'underwear' of the top edge of the undershirt you’re wearing, how much more of your underwear could I see?” I simply hadn't gone to that guard shack intending to pick up women, so I neither acknowledged the bait nor took it. She also stated she was curious about the reasons behind having a steering wheel on the opposite side of the car from the Brits, as well as driving on the right side of the road. But while the young lady inquiring about my “underwear” was unsuccessfully trying to get me to ask her out, another young female Brit was bold enough to interrupt and ask me out for a “date” first. Because of Beverly, I politely declined; that's just not where my heart and mind was at after the letter.

During that season in my life, I didn't hide my faith from fellow soldiers. Because I wasn't bashful about it, I was often subjected to much opposition and ridicule by my closest associates. The Maintenance Group would regularly plan activities and traditional drinking parties. Many from the shop took it personally when I expressed my resistance to their drinking and telling dirty jokes. (I can see now that I was spiritually immature.) So, when I explained my beliefs, I was usually misunderstood and came across as offensive. They retaliated with insults and anger. When others argued against my Christian position, I wasn’t well informed and was easily led off track. My open expressions of faith made me most unpopular with these airmen. On more than one occasion, a soldier wouldn't be content just telling me what they thought of me. Several were unhappy because, when they complained of our disagreements, the brass just told them to grow up. In attempts to try to get me into much more serious trouble, many did their best to provoke me into starting a physical fight. Through all these experiences, God taught me some important lessons. The guys were tough on me even though they truly liked me – except for my faith. When you speak up and live for Him, the culture is going to oppose you - an opposition that I keenly felt from all but one in my own maintenance group.

That one exception within the maintenance group was a friend by the name of Tutor. In stature he was a huge man - which matched his strong presence and personality. Tutor would often befriend and defend the small, the underdogs, the mistreated, and the unpopular. When he saw the sustained conflict between me and the other men of the shop, he suggested, *“Hey, Bowe. Let's go to town while these guys are having their party. If any of these guys bother you, I will beat the hell out of them even though I'm not a Christian like you are.”* So, we would take leave together and go sight-seeing; exploring the vast history that was all around us in the countryside of England.

For one of those trips, we decided to go to Cheltenham on a double-decker bus. We did do a lot of walking around together when we got there, but our different interests often saw us go our separate ways albeit in a friendly manner. My interest was in church history. Knowing that some of my forefathers came from Scottish ancestry, I also had an interest in the Scotts. Also, it was on the excursion to that small city that I purchased my very first suit. The fabric was light gray and so durable that I wore it for years

My interest in church history developed from hearing that the founders of the Methodist denomination, the Wesley Brothers, came from England and had established both churches and schools there. At the time, I did not know about C. H. Spurgeon whom many referred to as "the prince of preachers." His conversion took place at a Primitive Methodist Church meeting when he was just 15 years old. Speaking of that day, Spurgeon said, *"It was the happiest day of my life; I could have danced all the way home."* Following his conversion, he enrolled in school and was preaching at the London Tabernacle by the age of 19. He has written many Christian publications but had contempt for the state-run churches in England.

While I did not get to know all the men that were there with me in England, of those working in the shop and others out pulling security with me, I did get to know Pinky (Pinkerton), Red, Macintyre, Blain, Valdez, Willie, and Eels. Except for Senior Airmen Huff, James Mason, and Saint-Simon, who stayed in the shop for the entire TDY, I suppose most of us were loaned out of the Jet Shop at times for security and other details. Even though sometimes fairness didn't always seem to be a factor in the military, people were often recognized for their performance and promoted accordingly. Senior Airmen Huff and James were promoted to Sergeant soon after this TDY in recognition of their superior performance. Our top Master Sergeant's name was Harman, and the other was Sergeant First Class Jerome Krismer – a born-again Christian. They were excellent leaders and mechanics.

Before we knew it, our time in England was over. We had seen the other side of the world - what the folks back home called Europe, or simply referred to as "Over there". England is across a channel of water separated from France, Germany, Holland, Italy, Poland, and Russia. Getting packed and loaded was, as I stated before, part of our job in the SAC program. Our TDY group all worked well together getting the shop equipment and the extra engine accessories wrapped in protective coverings. We would be flying with the Navy again, but of course in the reverse order in which we came. The advance party, including the top sergeants, were flying out first but they had to leave one sergeant behind to keep all of us security and other loaned-out personnel in line. Some of the men could get pretty unruly over at the pub (commonly known as 'The Enlisted Club'). Others left women crying at the gates. Some told stories of being robbed, "losing" wedding bands, and all kinds of tall tales. Most of the men didn't share their exploits because after a while they were all the same.

Sammy Blain was among those who "sowed lots of wild oats" and was now praying for "crop failures". You see, Sammy had a line of possible, would-be mothers three miles long who tried all sorts of things to keep him from leaving. And like all the other women with "lovers", if they got your correct name and phone number and had become pregnant then the probability of an airman being in deep trouble was very possible. This was a great fear, especially for married men who hoped that they could get away with a fling while away from home on TDY. There were always men that had problems in town at the bars - usually centered around women and liquor. Many would return from TDY with fewer stripes than they'd had when we all left Mountain Home AFB. It was different with those of us who pulled security, we were all A/2C when we left, and we all came back that same rank.

Though some older airmen who were sharp did get early promotions, three men from the jet engine training group misbehaved and were reduced in rank. They had not been A/2C awfully long before they were reduced down to A/3C again. The Wyoming cowboy by the name of Landis and a Texan were arrested for helping themselves to gasoline at a construction site near Twin-Falls Idaho. Another airman's girlfriend became pregnant, and he would not marry the girl. So, the father of the girl did the correct thing to protect her and informed the Air Force Unit Commander of the situation. A lawsuit occurred with the airman paying double: He had to pay for support of the child and lost a stripe.

The Trip Back Home Over Big Water

Soon it was our turn to leave. The Navy R4D Skymaster taxied up into position. With the ladder down, a line formed to board. There were some mixed feelings among us about leaving. Some would like to have stayed and perhaps even marry a British girl - although probably not any of those that they had met at the gate shack. There are nice women everywhere if you know where to look, but those prowling around there were certainly not keepers.

Once aboard, the Navy stewardess again gave us brief instructions about the life vest: how to put it on and how to inflate it. Within 45 minutes we were out over the Atlantic Ocean. The next stop was the Azores with those same civilians feeding us out of soup bowls. I hoped that this time we would at least have nice bunks because we were going to be tired after that 11-hour flight. As before, our approach and landing at Lajes Field happened during the dark of night so we were again denied the high up vistas of that beautiful island; the home of the 65th Air Force Wing.

Before we knew it, it was time to get up and head out again. Why couldn't the sergeant let us sleep just a bit longer? After all, there was no war going on that we knew about. After turning in our bedding and checking out of the barracks, it was off to the mess hall to breakfast on good Portuguese bacon and eggs. We were all hoping that this time they would feed us on flat trays with real plates and use real cups for coffee, or maybe even just a linen napkin, but it was the same old standard, utilitarian military equipment. Even the best food doesn't taste that great without eye appeal and nice touches like regular tableware, but that isn't the Navy way - especially while traveling.

Once again, as we took off on the second leg of the trip towards our base back in the U.S., we loved watching the Navy stewardess showing us how to get into that survival vest (she put on a good show) and then settled in for the 10-hour flight to RCAF Station Shearwater, Nova Scotia. Even with the rest stop at Lajes putting us nearly halfway home from England, there were still many hours of flight time before we made landfall over Canada and a long flight on to Idaho. Soon after getting up in the air, anxiety got the better of me as I thought about going home. Where was home? Was my home now in Idaho where we were heading? Not really. The country around the base was just a much too desolate wilderness. I longed to get home, really home, where I could walk and hike the ridges and hollows of the Cumberland Plateau in Tennessee. That was where I wanted to be just now. I was due another 30-day leave (furlough) in August.

Flying over the Atlantic for hours was so boring with nothing to look at but the ocean and each other. A couple of card games were going, one up front and another in the back of this Navy R5D, when suddenly there was some action and everyone had to return to their seats. All that we knew was that the overheads flickered and the "fasten seat belt" light came on. Being a group of airplane maintenance mechanics, we all knew the inside workings of this plane better than the pilots did. What could be going wrong? We wondered if we had lost an engine or maybe just a generator or something. What could it be? We were all doing an analysis in our heads; meanwhile, there was no word from the flight deck - at least nothing given to us airmen. We all knew that this aircraft had four engines and in an emergency it could fly with just two of them; but, for the airplane to maintain lift with just two engines, significant weight would have to be jettisoned! That would mean tossing the luggage - anything heavy and not bolted down that we could lay our hands on might have to go. All aircraft have detailed manuals on both normal and emergency procedures. Whatever was happening, we hoped that this wouldn't turn into such an emergency way out over the ocean. Before long, we got the green light and everyone gave a cheer of relief.

Soon after, the crew gave us the estimated time of arrival at Shearwater. We just had to sit back and enjoy the ride for 40 more minutes until touch down. I know that I, for one, was looking forward to a nice bed, or floor, or anything actually solid that was not jittering, bouncing, or shaking. Hours and hours of looking at the back of seats and the tops of heads (some shiny and bald) is extremely boring. It was time to get on solid ground and get some rest.

Another cheer went up from the men as we touched down on the island of Nova Scotia. Being that much closer to our home base, with warm food and a bed that was perfectly still, relaxation began to ease into our minds and bodies. Conversations turned to what might have happened at the home base over the last several weeks of our absence. The married men talked about the problems their wives had written about: issues with the house, the car, and of course with the kids (the troubles they had in school and from missing their daddy). Many talked about having survived the brutal cold of the Idaho winter before we left. We had left Mountain Home in mid-spring just as the snow was departing and we would now arrive in the summer so things would be different.

One of my Christian friends, Bob Adair, had been listening in on all the married men's stories. Turning to me he asked, *"What about you Bowe?"* Well, I proceeded to tell him the story about receiving the "Dear John" letter. Bob, being the nice understanding man he was, replied, *"I heard some rumors about that. I saw her picture and thought that she was pretty, but you will be okay. I know you trust the Lord and He has something better for you."* Because he knew that he could trust me to keep things private, Bob also shared some personal things with me to let me know that he could relate to my situation concerning Beverly. He had a car and we had taken some trips to church together. Bob was much like Tutor in that he too was large, strong, and well-respected within the squadron.

After the second night of our trip, we turned in our bedding and again checked out of the temporary barracks for what we hoped would be the last time before getting home to Idaho. Ten weeks ago everything around here was wet and muddy, but now the walkways were dry as we made our way to breakfast and over to the BX for a few minutes of shopping. Nova Scotia was enjoying the change of seasons too.

At last, the Sergeant was calling for the troops to saddle up into the Air Force C-54 Skymaster that was taking us the rest of the way home. With only 30 minutes of our flight being over water this time, the rest of the voyage on our trip to the mainland would be less boring. Over the water time seemed to drag on and on but, with more interesting scenery below, time flew as we made our way across the northern Midwest and on to the Rockies. Everyone that could get to a window wanted to see as much as possible of the good old USA.

The closer we got to Idaho the more homesickness seemed to grow. While viewing the panorama of land from the air made it easy to guess where property lines lay on the straight rows of fences and trees, we could only guess what was in the fields (hay, corn, beans, wheat?); obviously this land was rich and productive. The further we went west, the larger the fields and cattle ranches became until it was hard to even tell where the boundaries were. Houses and barns appear so small from that altitude. As we passed over the Rockies, we saw a few snowcapped mountains here and there. So much beauty!

Finally, we were told to take our seats again and prepare to land at Mountain Home AFB. Wheels and flaps went down and not a moment too soon for any of us anxious airmen. We had successfully accomplished our temporary duty assignment in England and hadn't lost anyone. Everyone came back alive with all their arms and legs intact. In hindsight there were some losses. Because of wine and women, some had lost a few stripes but those could be replaced during the next promotion cycle. As both a unit and individuals, we were looking to settle down into the ordered routine that we had enjoyed before we departed in early spring for the TDY.

We must continue to stay vigilant as the "Russian Bear" is always on the prowl. About once a month, the group commander called us together and gave an intelligence report on the five-year plan of the Russians. Yet life had gone on while we were away overseas being vigilant: Babies were born, wives cheated on and/or deserted husbands, new autos were purchased, and old autos repaired. Nothing ever stays the same. I had lost my sweetheart (or at least had received a good-bye ultimatum) during that trip, prompting the statement 'Hell-O England, Good-Bye Heart!'

The year was 1955 when our unit returned from England, and we went to the Minor Repair Shop for duty and training. We had plenty of work for everyone in the unit, all of the engine change crews were so busy that their NCO asked for an extra man to fill his crew. I preferred this crew because the easy-going Senior NCO was a Christian and used better language than the Engine Change NCO who drank too much and told unhealthy jokes. For some reason that has escaped my recollection, it became necessary for me to go to the orderly room and while there I heard a TV news bulletin that an Air Force plane had crashed in South Carolina.

The Jet Engine Job Description

Ninety percent of my job was just removing and replacing standard parts that wear out quickly. An example of my workload would be as follows… The engines in those days had an ignition switch which was used to fire the sparkplug. It is okay for the pilot light to go out after the engine has been ignited since it is not necessary until the next start up. I had to carefully inspect the cables for damage or excessive corrosion, then remove the damaged, used ignition switches and replace them with new ones. Should the cable need to be replaced, then we did so too.

Another simple component to remove and replace was the fuel nozzle. Just like the carburetor works in your car engine, the nozzles spray fuel under pressure into a chamber; only a jet engine has about a dozen fuel nozzles. The nozzle has three main parts: A flat plate with screw holes, the crocked spray nozzle that goes inside, and a hose screwed to the outer side. The flat plate part has a dozen holes with a dozen screws secured by a safety wire. Cutting that wire and removing the screws is much easier than the reverse process. I had to pay attention to any damaged hoses or loose-fitting screw threads. The next component I also had to examine was the combustion chamber's outer liner. Generally, to replace that correctly, several chambers would need to be removed and replaced. The chambers are like stove pipes and serve a similar purpose. These chambers may vary in length from 2 inches up to 4 feet.

Our shop only worked on the J-47 engines that powered the B-47 bombers stationed at Mountain Home AFB (9th Bomb Wing, SAC Command). That engine had a flange at each end that the chamber rested or nested on, which was secured by clamps the size of large radiator hose clamps. Simple operations usually, we take off the fuel nozzles out of order, because the water-alcohol manifold had to be removed including hoses. The manifold is a large ring that is all around the engine near the fuel nozzle which sprays water mixed with alcohol into the combustion chamber for jet-assisted take off - thus adding thrust to the engine. The manifold was manufactured in two halves, then clamped together and secured for easy assembly.

One day when I was working on a B-47 bomber engine while it was secured on the tarmac, I noticed a fellow airman, short in stature, who I eventually found out to be Casey Collins from the electric shop, walking in my direction. I found this odd since most of us Jet Mech's would generally catch a truck to take us and our tools from the hangars to the flight line in the extremely hot weather. Yet here he came just a strolling with his tool belt on one shoulder and a spool of electrical wire on the other. I was high above on the engine stand sanding nicks out of the first stage of the compressor engine blades. After the test crew had adjusted the throttle linkage, I asked Airman Collins if he was here to assist me, but he replied that his orders pertained to a different area of the same aircraft.

This is a reference page on military jobs including the jet mechanic description. It reads as follows:

1. Maintain aircraft engines and their related systems including, cooling, fuel, oil, compression, combustion, turbine, gas turbine, compressor, exhaust, and propeller systems.
2. Conduct inspections on engine and engine-related systems.

3. Field-test and adjust engine components, fuel controls, pumps, and regulators.
4. Remove, repair, and replace turbine blades and combustion-chamber liners.
5. Preserve and de-preserve engines, engine accessories, and components.
6. Supervise engine work centers.
7. Supervise engine-change and minor repair crew.

George McIntire and I were doing electrical work. He and I spent two weeks trying to remove electrical wires from the wing of a B-47 Bomber. The wiring ran from the cabin the entire length of the wing and was strapped in at two-to-four-foot intervals. We had to lay on our stomachs and reach through small openings to unscrew the small bolts that held the brackets, which in turn held a cluster of wires.

Part of this was done at night for some reason; and because it was done at night, both of us were issued midnight chow passes. The maintenance crew would razz the electricians with sarcastic jests concerning their job. The NCOs would, in turn, come around and hound us by asking if we were finished yet. My guess was that the Generals had a schedule to keep, hence they would keep us out there for 12 hours a day. Being an electrician, Casey was a good sport about our playful harassment and would chime up and add to the jests just like everyone else did. Those of us in the Jet Engine Shop understood that we had an abundance of workers compared to the Electric Shop, which was short of qualified people. We would lovingly refer to the electricians as "spark chasers" and they in return would refer to us as either "stove pipe mechs" or "blow torch workers". Through this sarcastic banter between the shops, we developed a bond that took us through both the good and bad times. We in our shop never finished that job – I guess some General or other higher up decided we had other priorities.

Mt. Home First Baptist

Chaplin Major Dalton was someone who God used to play a major part in my life. He was the one who took me, along with several others, in the base van into Mt. Home to go to First Baptist Church. On April 23, 1955, Chaplin Dalton baptized us in the Name of the Father, the Son, and the Holy Spirit. We were inspired to spread our wings, go out, and do things we thought impossible the year before.

As base life started to get much more monotonous, God started to guide me into teaching and I was assigned the junior high class. The mix of individuals there were mostly 13 and 14 years old and, unbeknownst to me, were of that age where the girls were often more interested in the teacher than the lesson being taught. While this was a regular occurrence on Sunday mornings, it was distracting and annoying trying to focus the attention of girls too young for dating.

On our off time, we airmen were encouraged to go places aside from the base and its surroundings; even though those old tumble weeds were interesting to watch - especially when the wind picked them up to send them a rolling. There are several names for these plants, such as "Russian Thistle" or "Wind Witch". I found out that cattle eat them when they are green, but sheep do not. From the air, the tumbleweeds appear to be green pasture in the spring. One would see plenty of them on the way into the town and back. When driving at night and some come rolling across the highway, it is impossible not to be startled and apply the brakes.

In the fall of 1955, I rode to church with Bob Adair. I was enjoying the church worship for the first time since my seventh-grade Vacation Bible School days at Bethany Methodist Church, located just west of the Bowe Ridge. The Mt. Home Church had programs that the local Air Force people could attend, although they were considered too juvenile by some of the airmen. I did not go to First Baptist looking for a girlfriend, but there were pretty girls to look at, so I figured that was the place to be as a 21-year-old young man. Truly, I was having the time of my life in that church. It did my spirit good to get to know some of the other airmen that fellowshipped there; studying and commenting on the Scriptures together and seeing other young people have their own extraordinarily strong belief in Jesus.

The hayride was one of the first mixed-group activities that I remember attending. Although some of the other guys from the base were invited, most of them considered themselves too mature to go. There were three or four high school girls that went on it and one of them was a girl named Minnie. Although she was nice looking, I did not notice anything special about her at first. She had stood out to me some because of her involvement in several church activities: She sang in the choir, passed out literature, and read some of the church announcements. Minnie and I somehow ended up on the same side of the wagon which gave us quite a bit of time to talk. She wanted to confirm my name and the state I was from, so we chatted about what Tennessee was like compared to Idaho. We also had a lot of fun kicking and throwing hay in every direction. She was a large girl, not overweight, but thick, and just a big girl.

Later, Minnie approached me about going to the last football game of the season with her and one of her older, very attractive friends. Afterwards, she gave me her phone number and asked me to call her. Over the next year, we started getting to know and feel more comfortable around each other. The base bus was available to ride to and from there on a regular schedule during the week. Over the next three months, I would walk her the few blocks to her house after church; and did so several times before I kissed her on the cheek.

Minnie Dew's Family

One of the major tenants in the Christian religion is the doctrine of hospitality. Part of a church that cares about its people and members, Mrs. Dew did that well by inviting me to come to their house for Thanksgiving dinner. (If I didn't know any better, I would have thought she was playing matchmaker.) Wild turkey and elk were served for this unbelievably delicious meal. This was my first time trying elk and I found the taste to be close to that of beef. Minnie's Dad was very friendly toward me any time I went to their house. Mr. Fred was a hunter and that was the conversation at the dinner table. He and his son-in-law, David Buckman, talked about taking some pack horses up into the mountains to kill and dress out enough wild elk meat for the winter.

I had met David Buckman prior to this, knowing him as a civilian who worked in the Jet Shop. Due to his vulgar language, I tended to stay away from his crowd so as not to be influenced by him. I had no idea he was connected to the Dew family when I first met Minnie. Now that he was sitting at the table with us, I felt like I was eating crow when I realized he was most likely in ear shot when I made some remarks to some of the other young airmen about meeting a nice girl at church who was trying to get me to kiss her.

Both WWII vets, those two got a big laugh out of teasing me by informing me that my elbows were not allowed on the table if I had not been overseas. It seems that Fred, like David, had the "one up" on me because his daughter, Mrs. Buckman, had an office job there on the Air Force Base and had checked my records to make sure there wasn't a wife or child back in the state of Tennessee. They said my military record showed no bad marks and that I was sending an allotment to my parents. It was apparent that Minnie was well protected by her parents, sister, and two brothers, and they wanted to keep her a sweet, innocent little girl for as long as possible. Knowing I was older and more experienced, they certainly could not be blamed for that.

Plane Crash in Charleston, SC - 9/25/55

I made the remark that my brother Bud was a crewman on one of the C-119 aircrafts based in Charleston. When I heard the news report that day in the infirmary, I tried to dismiss the thought that it might have been his plane that crashed in a community and burned. As I knew pilots always try to turn their crippled aircraft away from residential housing, something had obviously gone very wrong. The account as detailed by the Charleston Associated Press stated, "Shortly after a nighttime take-off from Charleston Airport, during the ascent one of the engines caught fire and exploded. The airplane lost height and crashed in flames into several houses located in a residential area some 1.7 miles from the airfield. Six of the Air Force crew were seriously injured and five were killed; while four people in the houses were killed as well."

Turned out, A/2C Bud H. Bowe *was* one of the survivors taken from the C-119 crash site to the hospital for examination and treatment. Of course, there were many injuries, but Bud's most serious was a concussion and deep shoulder bruises. My brother remained in the hospital for some weeks so the medical section could determine the extent of his injuries beyond the concussion. It would be 90 days before he could fly again after physical and psychological therapy.

As he recollected the frightening experience… Bud was seated in his customary location for all takeoffs and landings - safely strapped in the designated rear portion of the cabin along with four other airmen. In the aftermath, while the main portion of the boxcar was still burning, Bud said he remembers coming to and can't recall if he unstrapped himself or if perhaps the strapping had been torn loose upon impact. While barely conscious, he remembers crawling up a few feet to a rear opening where he rolled out and then fell about 15 feet to the ground.

The back of the plane was resting on structural debris or what was left of the fuselage, so the rear was projected up into the air. Two other crewmen were assisting each other, and another crewman carried Bud on his shoulder away from the burning, smoldering wreckage before collapsing onto the street himself. Two seated in the rear, including the Airman seated beside him, and the entire cockpit crew (Captain Jones the pilot, Lieutenant Jefferson the co-pilot, and Samuels the Flight Engineer) were all killed on impact.

It was fortunate that no paratroopers or special forces attachments were aboard. The official flight plan was for a training exercise that would have included all the crew members, but this was only a training mission from the home base to Denver, Colorado. When they were overseas, flying from an unknown base on intelligence-gathering missions, the job was quite different.

Bureau of Aircraft Accidents Archives

CRASH OF A KAISER-FRAZER C-119F FLYING BOXCAR IN CHARLESTON: 9 KILLED

Date & Time:	Aug 23, 1955 at 0200 LT
Type of aircraft:	Fairchild C-119 Flying Boxcar
Operator:	US AF
Registration:	51-8165
Flight Phase:	Takeoff (climb)
Flight Type:	Training
Survivors:	Yes
Site:	City
Schedule:	Charleston - Charleston
MSN:	168
YOM:	1952
Location:	Charleston, South Carolina
Country:	United States of America
Region:	North America
Crew on board:	11
Crew fatalities:	5
Pax on board:	0
Pax fatalities:	0
Other fatalities:	4
Total fatalities:	9
Circumstances:	Shortly after a night takeoff at Charleston Airport, while climbing, one of the engine caught fire and exploded. The airplane lost height and crashed in flames onto several houses located in a residential area, some 1,7 mile from the airfield. Six crew members were seriously injured while five others were killed. Four people in the houses were killed as well.
Probable cause:	Engine explosion during initial climb for unknown reason.

A2C Buddy Bowe, the middle son of the family, was an airman on the C-119 like the ones

This picture of the C-119 Boxcar with its clamshell open to deliver supplies to a truck reveals just how large the inside is.

Pictured below: The burning wreckage of the C-119 crash from Charleston AFB that my brother Buddy was injured in.

Cold War Ops

Bud described his adventures training on the C-119 "Flying Boxcar" as both intense and dangerous. These planes had been specially modified to retrieve military weather-type balloons in mid-air for a program called or referred to as 'The Dragnet'. These surveillance devices had been strategically launched so that the jet-stream carried them across enemy Chinese or Soviet territories in hopes of intercepting signals and/or pictures of something of importance to the Armed Forces. Once the wind finally blew them back into safe airspace, it was the job of my brother and others like him to capture them so the data they had collected could be processed.

Basically, the 'clamshell' or cargo end in the rear of the plane would open up and men would work in tandem to extend the device which caught the balloons. Bud was a 'Pole Handler' - responsible for extending or retracting and positioning the fiberglass poles of one side. The winches and hydraulic extenders had to be timed together, with each man attaching or detaching the hooks and cables that formed the butterfly-type net when extended. After extending the device, it hinged downward in the retrieving position. Multiple training exercises were performed in preparation for catching these balloons in mid-air. Their huge plane would dive down close enough to pick up a barrel off the ground. Other exercises were conducted over lakes or ponds whereby the aircraft would repeat the dive and retrieve the barrels from the water.

While stationed in Adak, Alaska, my brother was involved in an actual attempt to catch one of these balloons in flight. The planes worked in units of three flying the same mission. One air crew spotted a balloon and did a dive in full readiness to pluck it from the sky.

As usual, because catches were rare in most cases, they missed. Next, Bud's crew attempted the capture; but it was one elusive balloon. Another one of the other aircrafts flew into retrieving position, dove on the balloon, and missed also. For this mission, the 4th time was the charm and my brother's crew were the ones successful at catching the balloon with its critical intel! The team of happy warriors received some awards and benefits for the catch. [Eventually the program had to be terminated when the Russians exposed the balloons and their black spying boxes on some newscast.]

Buddy flew another year with the 119's and cross-trained as a Loadmaster on the C-124 before he got out of the service. In hopes of becoming a surveyor, he and his wife returned home to Tennessee and lived with our parents. When that didn't work out, he decided to re-enlist. Even though he hadn't even been out for a year, unfortunately it had been just long enough to lose his pay grade and benefits.

The Pilots Thoughts About the B-47 Bomber – "Great!"

When delivery of the sweptwing supersonic B-47 aircraft began arriving in 1951, most crews were excited about getting to fly the new bomber as it was so much like the jetfighters of that era. The B-36 'Peacemaker' was slow and cumbersome in comparison. The B-47 was so fast it set performance records in the early days with ease. The aircraft handled well in flight with a "fighter-like" touch to the controls. The large bubble canopy enhanced the feel for the pilots via improved vision, but the design would cause unusual temperature control issues for them as well.

The 3-man crew consisted of the Aircraft Commander, Pilot, Co-Pilot Navigator/Crew Chief. As a member of the ground crew, the Crew Chief was eligible to receive an extra $50 flight pay per month (because the plane was such a maintenance hog) if he flew at least the required 4 hours. In May of 1951, the first B-47's in the United States Strategic Air Command (SAC) were assigned to the 306th Medium Bombardment Wing based at MacDill AFB, Florida. When fully operational (which took until the year 1953), the B-47 Bombardment Wings were divided among the 2nd, 8th, and 15th divisions of the Air Force. Most of the Fifteenth's aircraft were stationed in the Northwest. A squadron of 15 aircraft required 77 officers and 99 airmen to maintain.

The B-52 Heavy Bombardment Squadron also had 15 aircraft in each of the three squadrons or 45 aircraft in each Wing with 173 officers and 243 airmen. The combat crew consisted of the Commander Pilots, Bombardier, Navigator, Electronics Counter-measure Officers, and a Tail Gunner. The heavy B-36 bombers that preceded the B-52 Wings also had 3 squadrons in each Wing but only 10 aircraft in each squadron. In March of 1961, President Kennedy directed the phase-out of the B-47's. This was temporarily halted because of the Berlin Crisis of 1961-62, but gradually they began delivery to storage.

TDY #2: SAC Bombing Competition - 1956

And then there was our 9th Field Maintenance Squadron's involvement in the Bomb Comp (SAC Bombing and Navigation Competition) Mission to Davis Monthan AFB in Arizona. The actual details of the competition were excitedly discussed in the enlisted ranks because they were affected by the outcomes – either promotion or possible demotion. To us maintenance crews, however, the specifics regarding targets, wings, or groups involved were irrelevant. We knew what our job was going to be - remove and replace parts for whoever.

As a mechanic, I had been assigned a toolbox at Mt. Home which remained as part of my gear for the rest of my time at that base. It stayed with me in the home shop and accompanied me on our strategic deployments. One day, the shop supervisors directed us to check out our toolboxes and pack travel clothes for a few days. The first part of this trip was a bit frightening. We flew out on a two engine C-47, which was the type of aircraft that was used to drop paratroopers in France during the Normandy invasion in World War II.

That plane was loaded to capacity! From the time we started down the runway until we landed at Tucson, the aircraft was shaking and vibrating as if it were coming apart. We flew *through* high mountain passes, which meant the mountains were higher than we were flying! I can't tell you what went through my mind when I looked out the windows and saw a mountain peak on each side. My memory recalled the movies, seeing the news reel, or live films about 'Flying the Hump' and the Burma Road events in World War II that supplied the American and Allied troops behind enemy lines. That is very similar to the view I saw as we flew out for that mission. We all looked at each other as though seeking reassurance. The older Sergeants tried to look calm and smiled while making jests like, *"This is nothing. We've been here before!"* Although at times many of us had reasons to doubt if we were going to make it to our destination, we did manage to arrive at Davis Monthan AFB in Tucson, Arizona, in one piece.

We spent a couple days performing our normal maintenance tasks and some other duties; one of which was honing rotor blades. This job entailed having my head and upper body positioned in the front of the engine while taking some nicks out of the intake blades via an Arkansas stone. As these bombers taxi to the runway, they suck up many small stones and other debris which tend to nick the engine. I've spent many hours honing rotor blades to prevent larger cracks from forming. (We were instructed not to use a file on them, but they gave us small Swiss files?!)

Casey Collins

Airman Second Class Casey Collins was a small man with a big heart. He was what some of us called a "spark chaser" (i.e., being highly skilled at repairing the complex electrical systems of our B-47 bombers). I had known Casey for some time as he worked in the electrical shop near our jet engine shop. He liked to tease me about being a "stove pipe mechanic" or a "blow torch worker" in reference to the exhausts of the engines, but it was all in good fun as the joking always went both ways. Once, when Airman McIntire and I were on a night job replacing a wire for the marker light bulb at the end of the wingtip, we got a couple of midnight chow passes and happened to meet up with Casey while eating. We ribbed him about us having to do his job. We well knew that the electric shop was short-staffed and had more work than they could ever do, so we were glad to pitch in and help with a simple wire repair.

Collins was a good man and a hard worker. It seems that good electricians were always scarce in the military - especially electricians who knew and understood the complex circuits, lights, diodes, capacitors, rheostats, potentiometers, and switches of a B-47 bomber. Perhaps that scarcity was because electricity can be somewhat of a troubling mystery for many people. First, they don't like the sudden surprise and tingly feeling of getting shocked; and second, electricity seems difficult to understand, much less contain and control unless you understand the basics of voltage and amperage. So, not many recruits elected to go to electrical school. Those that did go, and became proficient, often elected to transition into civilian life at the first opportunity because of higher paying jobs.

Casey was a good electrician. He was a small man and could often be seen lugging his tool belt and parts out to the flight line to carry out his assignments. Even though there were drivers assigned to transport mechanics and electricians to their job sites, Casey would most often turn down the ride and just walk to his.

One day when I was out on the flight line working high up on an engine stand honing some nicks out of a turbine blade, I watched as Casey trudged out of the shop with his tool belt: this time carrying a roll of wire on his shoulder. When he got near, I asked him why he didn't catch a ride with the transportation guys. He laughed and said that he just didn't want to bother them. Besides, he had seen them careening around the big jets taking chances that worried him; so, he didn't trust their driving. It was safer to walk, and he had the time and needed the exercise. I could certainly understand that because I had seen those cut-ups from motor pool myself and had to shake my head and wonder who in the world taught them how to drive. They drove like they were in a demolition derby. (And why were they allowed to continue driving when they seemed so careless?)

Not long after my conversation with Casey, he had apparently worked several very long days chasing down and repairing electrical shorts, repairing equipment, and doing what he did so efficiently and well. Nobody knows for sure exactly what happened, but he had probably taken a break and laid down in the shade of the wing of a B-47 bomber. In the process of getting a much-needed nap, he apparently snoozed on past sundown into the evening, which is when the tragic accident happened. Being in the Arizona desert with truly very few city lights, the night got very dark.

A couple of guys from transportation were out racing around in a pickup truck and, in one of their turns, they cut in much too close to the airplane. In their carelessness, they simply didn't see Casey sleeping there under the wing and ran over him. He did not survive his injuries.

When we found out about the accident, we were told that the driver might have been blinded by the bright flame of his cigarette lighter so he couldn't see Casey lying there. That didn't make sense to us because we knew that cigarettes and lighters were not allowed on the flight line because of fire danger around so much flammable jet fuel. Many investigations took place about what, when, and where was the supervisor or officer in charge etc. Lots of questions needed to be answered, such as: Why was a truck driving so close to the aircraft? What was Collins doing under the wing, in what should have been the safest location from auto traffic, at that time of night? What was the reason that the airmen decided to violate directives put in place to prevent planes from exploding? Was that just a lame excuse? Who knows? We heard other versions of the story floating around in an attempt to explain the accident. None were satisfactory. None were plausible.

Accidents are accidents and always happen because of choices that people make. The bottom line was that Casey Collins, our friend, was now gone. What were his parents told when they were notified by the Red Cross or by military brass? Were they told that Casey Collins died honorably while in the line of duty? Doubtless there are other similar stories because, even in the military, accidents happen and frequently cause injury and death.

As for the Bomb Comps, we never knew who won, lost, or was disqualified. We, the 9th Field Maintenance Squadron, must have never won anything - it seems the winner usually got a trophy. We were all packed up, awaiting an old C-119 after being bussed to the flight line. This aircraft has a large body in the middle with what looks like the ends of two fighter jets making up its split tail. These were incredibly good aircraft.

Manufactured by Fairchild Corp in 1947 when it made its initial flight, it was used much like the C-47 for air drops. In addition, this aircraft could haul cargo and equipment such as jeeps, trucks, howitzers, and other large cargo. The C-119 was used extensively throughout the Korean war. The whole 9th Squadron was anxious to get back to Mountain Home. We didn't know this pilot and commander that briefed us with such stern warnings and cautions like, "If 'X' happens, here is what you do." His manner made us wonder if he was somewhat afraid of the airplane himself or was he seriously cautioning us? Since we had never had anyone speak to us that way before, his demeanor made us uneasy.

We loaded up and went down the runway with much more noise, shaking, trembling, and bouncing than any of us were used to. I was praying in the same manner as I had on my first airplane ride at Lackland

AFB; wondering if this was the day to "cash it in". We again repeated the experience of going through the mountain pass with peaks higher than we were flying. It was nice when we touched down at home base.

Next of course was the funeral for Casey. The Electric Shop crew were allowed to be the pallbearers. Those that I knew from the Bomb Comp group talked about how much the family appreciated and loved the men who escorted Casey's body home. Obviously, the family was grief stricken and wanted some answers. None of Casey's friends were with him at the time of the accident. The family wanted his best friend to stay with them, so they invited him to come and live with them after he finished his time in the Air Force.

Speaking of going places and doing things, the Air Force gave us a free pass on weekends if we showed on Monday at our workstation. Sammy Blain seems to have met a "friend" in Boise, Idaho with one of these weekend passes. He was going the 40 miles to visit her regularly and, of course, still drinking heavily even though he had to borrow money and would "forget" to pay it back at the end of the month. Home is always the best place to take rest and relaxation for a weary soldier, and it had been 20 months since I had last seen it. I desperately needed a break from the monotony of military life's day to day activities. For certain, I really missed the green vegetation of the hills and valleys of Tennessee as I was tired of my Idaho desert surroundings.

Train Ride to Cincinnati

Cookeville… I kept wondering if things had changed since being gone for twenty long months. "What changes were in store at home with all the Bowe boys in the Air Force?" I pondered, and are my sisters dating now? Ed Keaton was also an enlisted airman by this time, as were Clyde Nickels and Charles Marlow. My mind thinks of the revolutionary wave of military patriotism that ran through our community. The first individuals were from the Gentry or perhaps Rocksprings area. One may have been from the Pulley family. There were stories about him and Jose Salazar; the way they wore those uniforms so well, walked so neat with shoulders back and shined boots, receiving admiring looks and whistles from the girls. Home is where the heart is. Whatever all the changes may be, whether it is to my liking or not, I sure was anxious to get there. There was no doubt in my mind that I would receive a good night's rest when I made it home. Sleep did not come easily on the first leg of my journey from Idaho to Cincinnati; it was like I could feel every rattle-rattle, jerk-jerk that train made.

That ride from Mountain Home to Cincinnati was going to take two weary days. We would be crossing the Idaho desert for the first couple of hours or so with two stops before reaching Twin Falls and then Pocatello. One of the stops was at Glenn's Ferry where it was rumored that some years ago a fisherman had caught a sturgeon so large that they had to summon a farmer with his tractor in order to come pull the fish up onto the bank of the Snake River. Here, we changed cars to one that had a scenic upper deck and a group of what I believed were female college students came aboard. Some of the ladies, seemingly attracted by my uniform, began asking questions such as what were soldiers doing here in Idaho or was I just passing through. When I told them that I was an airman, they supposed me to be a pilot.

One of the ladies seemed a bit more aggressive and we fell into some interesting conversations over coffee together in the food car. Later, at her suggestion, we moved up to the scenic deck to take in the breathtaking view. Now her interest seemed to shift from me to the scenery; it was unbelievably beautiful as the sun was setting over the mountains. We seemed to agree on a variety of subjects, nothing important just small talk. Once the sun went down, the silence between us grew right along with the darkness and I began to feel self-conscious as we both realized we were alone. Unsure how to proceed, I offered to purchase her something to eat or drink but she refused both and just kept looking away from me. Taking a chance that maybe she was lonely too, I slipped my arm around her. She said, *"Excuse me"* and left. I was embarrassed now, left sitting there in the scenic car all alone. There was always a 50/50 chance that my actions could spoil a perfect evening, but it was worth trying anyway. She was older and smarter than I, perhaps in her 30's. The most that I could think to do from that point was just to go to sleep and look forward to the next day on the train.

Descending the stairs, I found a window seat as the train stopped and discovered we were now in Wyoming. As we rode through those mountain passes and tunnels, I sat there alone trying not to breathe in too deeply; fearing that the diesel fumes and smoke belching from the train engines would make me sick and add to my unhappy thoughts. Blaming myself for not being the gentleman I was supposed to be,

I wondered just who those women had been anyway; were they some Catholic order of nuns or perhaps nurses or Amish? They had bombarded me with questions, and I had not thought to inquire about them. (They had all been dressed similarly in modest, grey with black trim outfits and not wearing makeup.) After taking many small cat naps between the rough places on the tracks and wishing I could be in one of the sleeper cars, I finally fell asleep despite all the jerking and shaking as the train rounded the curves and rumbled through the steep and hewed-out passages of the Rocky Mountains.

The next morning after breakfast I went roaming through the cars. I happened upon and met some other service members while on the prowl for another unaccompanied female. As I walked the isle, a young lady passed me by going in the other direction. I had not seen her the day before, so I deduced that she must have boarded in Wyoming or Nebraska. After walking through another passenger car, I returned to where my seat was and again this same lady and I met in the isle where we exchanged pleasantries but kept moving. After a few steps I turned to steal a glance, for a closer look at her walk and her build, and to my surprise she had also turned for some reason. I didn't know what the next step should be since it seemed that I had been caught checking her out.

There was an empty seat in that car, so I decided why not take it and see where she is going to sit. After some time, she went to the restroom or to the food car. Eventually she walked by my seat, and I noticed she was wearing a ring. That answered that for me, so I started to see if there was another person that might be interested in talking about the ride thus far when I noticed a couple service members having a conversation and decided they needed another military man's view on the subject matter. We quickly became loud which drew a couple of ladies into the conversation, but we were also told to hold our voices down. The young lady wearing the ring also joined in. It became evident that we needed more room, so our large group of people moved to the back of the car. Since I was now close to the young lady, I inquired about her husband or fiancée. She replied, *"Oh I just wear the ring to keep the undesirables away."* She then took the ring off and placed it in her pocket.

Time passed quickly after that as she and I took seats together and started chit-chatting. We learned each other's names and destinations, background, and plans. Her name was Helen, and she was returning from a round trip to Pennsylvania. We just sat and looked out the windows, making remarks about the scenery along the tracks until it was getting time to eat, and I offered to pay for her lunch. We ordered our meals and drinks and sat looking at each other closely, smiling in admiration. Finishing the lunch I could hardly afford, we found a window seat and gazed at each other in almost a penetrating fashion as we began a quick review of our lives and discussed our likes and dislikes in politics, family, work ethics, dream homes and places we preferred. I had never discovered so much about women and myself as we opened up to each other! We were both country (she from Pennsylvania and me from Tennessee) but it was surprising how similar our backgrounds really were. Although Helen had not worked in tobacco or with mules like I had, she did the housework and cooking as my mother and sisters had done while my brothers and I worked in the fields with dad. She related growing up with her brothers and a dad who left home for war.

To me, Helen was the most beautiful woman I had ever touched. Tall for a woman (around 5'7") and thin but not too thin (like a strong-looking farm girl) with long brown hair that fell over her shoulders like a shawl. She was so attractive with those large, beautiful blue eyes and such a big smile. We became so close in that short time together. Before reaching Cincinnati, my heart just melted! Not being able to touch and smell a woman for such an extended time, she was so soft and tender and smelled so good. In my mind, I was like, "Mom, she's so pretty! Can't I keep her?"

Helen and I went for a walk at the station in Chicago and did some window shopping nearby. We met a street salesman who tried to sell me a wristwatch. Suddenly there appeared a city policeman who spoke very sternly to him, urging him to move along. The officer also gave us a warning about stopping and listening to his type, who had cornered us in a store front. The officer knew the man as a thief and swindler and said, *"If you would have taken out your wallet to pay for that watch, the wallet and your money would have been gone."* We thanked the officer and moved a block closer to the train station, having learned well a short lesson about street life.

Helen was too willing to follow and agreed with all my suggestions. We both were en route but delayed at the train station. I was going south, and she was going to continue east. In our conversations I had discovered this fact about Helen; she was a country girl from the mining part of the country and, like me, had been poorly educated. When I found out where she was from, I am sure that I told her about my Aunt Willene living in Pennsylvania, and even the story about how she had met David Cope, who was a famous person from there. Helen was in the same situation as myself - she was looking for her knight in shining armor and I was looking for my Cinderella who lost her slipper. We were so attracted to each other. I looked at her with admiration and she looked at me with those beautiful blue eyes full of happy expectation.

The time to part ways came more quickly than either one of us cared for, but I had to board the south bound bus for Knoxville, Tennessee. There was a question rattling around in my mind. If I would have said, "Let's buy you a ticket to Tennessee and you can put that ring back on," would she have agreed to do that without much hesitation because we were letting our hearts lead instead of our heads and good common sense? We did exchange phone numbers and addresses.

Those two days together were beautiful and nice to remember. That trip from Cincinnati to Tennessee was such a long ride; and now my imagination ran wild, daydreaming of the possibilities that could or might happen later with Helen. I pondered in my heart and mind how real was all of that? Did I do and say the proper things? What was meaningful that we talked about: Religion, the Bible, our own personal beliefs? The thought came to me, "I can write about those things later, those things are important in a caring relationship." What did I really know about relationships? Before I knew it, I was boarding the bus to Cookeville. I attempted to call her up and pursue her more, even long distance, however she was not a writer. We gave up on each other within 6 months.

Back at Home: Leave 1956

Mothers are always mothers, and my mom was no different; no matter how old you get, you were still her child. She worried that, with me now being a military man, I would be more open to the influences of the ways of the world and she believed some counseling was in order. We hadn't seen or talked to each other in almost two years and she had a thousand questions for me. Before I could answer one, she would ask another; it was as if she had been saving them up! Finally, I told her that I believed there was something for her in my B-4 bag. (A B-4 bag is a special, compartmentalized suitcase you purchased at the BX.) As soon as I would take items out, my sisters wanted to touch and handle them.

Next morning, thanks to my military schedule, I was out of the bed early which meant I got to see more of my dad who's been an early riser all his life. He would always wake up with something on his mind, fussing and carrying on about something or another each morning. I found I was still accustomed to his routine, and commotions, etc. His antics made me laugh and I asked about his dogs and how the hunting had been in the area. Other than that I don't remember much, except the smile showing that he was glad to see me.

And then there was Mother's home cooked breakfast! It was impossible to ignore those open-face biscuits and her delicious gravy! Growing up, if I were lucky, I got a pork chop, too. For my first morning home in so long, she made me such a special breakfast - not the usual every morning event. Words weren't needed to tell mom how much I had missed that gravy in the mornings; she could tell by the way I devoured everything in short order! I told her about the Air Force's S.O.S. breakfast, which was basically hamburger creamed gravy served alongside of toast; and recounted how often, along with beans, these were on the morning menu. This prompted her to ask about my pulling KP two weeks in a row while at Chanute, AFB, which she remembered from the letters that I had sent her. In our conversations about England, I told them that the movies had it right concerning what to expect if they were ever to go to there. They have rock walls, cobblestone streets (which of course they drive on the left side of), they dressed differently and, while they spoke English, there were certainly different terms for things like the bathroom.

My sister Shirley, who was seven, was zipping around as if she owned the place. Carol was fourteen and dreamed of a different life since she didn't like school or home then. Peggy was dating and it was expected she would most likely be out the door very shortly. After breakfast mom wanted to sit and talk more as she was hungry for a different perspective than her daily hometown experiences. However, I remember wanting to get moving as soon as possible in search of a suitable car for the right price. (Part of the money I had sent home during my time in the Air Force had been set aside for this purpose.) Dad was working in a small corn crop and engaged in some fox hunting every time he had the chance. Even though farming was not that interesting to me, I do remember spending some time with dad before setting out on my car-finding mission.

Another thought as I went into town was to try to find some of my schoolmates and friends. To my disappointment, all of the old gang was gone. We, my brothers and I, didn't really have any close friends except for the gang that had gone into the Air Force; including Red Neil. During my exploration after two years away, I did find one gal still working around our hometown that I had become good friends with while in school by the name of Bell Brown. There was never any romance between us, just good Christian friendship, and she volunteered to keep me company while I went car shopping.

We borrowed a vehicle to drive as we looked for that special, 'new-to-me' car that was going to transport me those nineteen hundred or so miles back to Mountain Home, Idaho. While I believe that I was taken with one of the first cars I drove, Miss Brown and I drove to Nashville, Lebanon, Sparta, and Carthage, Tennessee, as I wanted to shop around just in case. After looking around but not finding anything that I liked more, that was the one that I purchased. In addition to making the dealer reduce the price, I also made him give me a warranty that insured the car would get me to Idaho or else. The warranty as written stated something to the effect of: Should the car not make the trip, they would pick up the car wherever it became disabled, they would return my money, and I would owe no further payments.

Of course, upon returning home it was necessary for me to take everyone for a ride in my new 1952 Ford. Mom was the only one who seemed unhappy. Because she was still hungry to hear more of my experiences outside of our town, she thought I should park the car and sit and talk to her. It was quite understandable for my mom to feel that way since she hadn't seen her son for over twenty months and didn't know when and if she would ever see me again.

Driving from Tennessee to Idaho

Wow, a new car in my very own name! That was a dream come true since Ms. Detroit and I had become stalled in the woods two years ago. That ordeal still haunts me somewhat since I never received an explanation for why our relationship ended. Just when I think it is all forgotten, an echo comes in the night no matter where I am to remind me of love lost.

As seems to be the case with anything, problems arise, and temptation comes with the territory. A dear friend who I hadn't seen in quite some time wanted a favor. He had scheduled a date and his car had broken down. On top of that, then his date wouldn't go unless her friend went also. Since the reputation of his lady friend had preceded her, I declined the blind date fearing that her young friend was of the same caliber. I believe this could have been one of those traps the Lord saved me from.

A military furlough is the shortest 30-day month. The process of signing out counts as one, then minus three days for the train and bus ride home and the travel time it takes for the return trip, that only leaves eighteen or nineteen days at home. I am not complaining; one of the nice things about being in the Air Force is the travel - and they do pay each month. It still is the American dream! What other country has more to offer?

Well, I had come home to Tennessee to see the folks and to purchase a car. Vehicle ownership carries responsibilities. My 1952 Ford checked out okay mechanically. The tires were only slightly worn and the dealership performed a full-service job: oil change, fluids topped off, etc. They advised me to secure insurance in case of an accident. We checked with AAA for the best route from there back to my base. Mom and the girls cleaned and washed the car in preparation for the trip. She also cooked up some chicken and packed me a gallon of tomato juice. I guess we teared up a bit as we prayed before I left. They asked, *"When can you come back again?"* Unfortunately, I had to tell them it would be 12 months before I could even apply for another furlough.

Following the route AAA had planned for me, I left early in the morning headed west toward Nashville and on through Memphis, Tennessee into Arkansas and Northwest. Butterflies were all up in my stomach as I was just a bit fearful that the car might not make the trip. It took a lot of faith and trust in Jesus to get me back in one piece, in this new-to-me machinery. Somewhere in Missouri, I stopped beside the road overlooking a nice green hayfield. I was hungry and that chicken and tomato juice sure were refreshing.

Back on the road again, the car was running well and getting good gas mileage. Gas was not very expensive in the year 1956 (about 30 cents a gallon), which was a good thing since I didn't have much money after purchasing the car. I hadn't calculated for taxes and insurance, so money was short, and every dollar had to count in order for me to reach Idaho. With only a few dollars left, I remembered I did have a safety net: Service members can take their orders to and get paid at any base.

Driving through western Missouri, a soldier was hitch-hiking out on the open highway. He was in full dress uniform and had both a duffle and a carry-on bag with him. Responsibility is a funny thing sometimes and owning any kind of transportation entails certain responsibilities of its own. One must consider the dangers that exist such as a potential crash, hands getting caught in doors and windows, blowing a tire and losing control, etcetera - and even more so with a passenger. Nevertheless, without much hesitation or thought, I decided this person needed a lift in the worst way. A precept in the Bible is kind of summed up by a song written by Glen Campbell, whose lyrics go something like: "If you see your brother standing by the road with a heavy load, stop and say, 'You're going the wrong way.' So show a little kindness." As I slowed to a stop, he picked up his bags. Getting out of the car and opening the lid to the trunk, I said, *"Buddy, where in the world are you going?"* He replied, *"About as far west as a person can go."* I told him I was bound for Idaho, to which he replied, *"This is my lucky day!"* If you have a smidgen of faith, you can see God in this. Two people have great needs: One person needs help driving this long distance, the other needs to get to a distant place, and we both need to fulfill our responsibility to our master (the U.S. Government) by getting to our place of useful service. A verse the Apostle Paul wrote came to mind, "Some planted, some watered, and God gave the increase." 1 Cor 3:6

After learning his name and listening to his story about how he didn't have enough money for the bus ticket, etc., I told him getting along on this trip will take some understanding and informed him of a couple of simple rules. (The way I figured it, no soldier should have trouble following rules.) Rule number one: No smoking in the car and no profanity or dirty jokes - there are plenty of clean ones. I needed to stretch my legs regularly and rest my eyes, and since Joe did have a valid driver's license, I said he could help drive under one more condition: Treat the car with the utmost respect since it is our only hope of getting to our destination. *"Should those simple requests be too difficult,"* I said, *"then you are free to ride your thumb again. You were able to find this ride, so surely there would be another."*

I was hoping to make it to Grand Island, Nebraska, the first day. We were getting along on the highway and with each other very well. Traffic was moderate but the sun was in our eyes. Joe finally asked to drive a short distance to the next gas stop. I watched every move as he pulled from the curb onto the main highway and found he had all the confidence and skill needed to drive my car. We agreed to drive as late as possible, then take what sleep we could manage to get in the car to keep moving. Upon waking up from a nap at one point, I noticed that Joe was driving 20 miles per hour over the speed limit. Knowing that my 52' Ford could not take that abuse, I slowed him down and started to pay more attention to how he drove my vehicle. On and on, we drove and drove until the car needed gas again. Joe felt confident that we were going all the way together, so he chipped in for a tank of gas. (In those days, eight or ten dollars would fill the tank up.) We drove into the wee hours of the morning until both of us were so very tired that we had no trouble falling asleep even in the car. We parked at a service station in Nebraska someplace and I shared a piece of my mother's chicken and the tomato juice with Joe. That wasn't much of a breakfast, but we intended to find a nicer stopping place soon. The rest of our journey was long but uneventful. Joe was in the Army and, although I took him 100 miles out of my way to where he was stationed in Washington, having the two of us driving made it to where I still got back to my base on time.

Mt. Home Again – Autumn 1956

Well surprise, surprise, that 52 Ford made it the entire 1900 miles! The car wasn't running as good as when I left Tennessee due to Joe's driving though. If a person is mechanically inclined at all, they will notice different sounds in the motor and the roar or hum of the tires on the road. I still loved the car, but it was using oil and smoking slightly. Now I looked forward to getting back into my church activities at First Baptist Mountain Home! The 10-mile drive on Sunday morning would feel great! I volunteered to sing in the choir and teach a Sunday school class.

On the base, things weren't much different. Several of us A/2C's that had pulled security in England were scheduled to do TOC's (Tech Order Compliances) on the six engines located on the B-47's outboard wings; some were warnings, most were electrical or mechanical changes to prevent problems. After the towing crews pull the aircraft into the hangars, we take the engine stands and proceed to remove the cowling (outer metal covering) from the engine. We then locate the electrical connections and remove them, and then find and remove the bolts and a couple of screws that secure the alternator onto the engine. Next, with some hydraulic lifts and two airmen, we removed the alternator from the engine and its housing so we could exchange it out or perform modifications.

Minnie Dew

There is an off chance that Minnie could have been only 15 years old when we went on that hayride the previous year. She was in high school and did not want to talk about that fact very much. Wanting to be more mature than her age, she could easily have passed for nineteen or older if it weren't for her very pretty babyface. Despite the age gap, Minnie and I had connected and were consistent in getting together after church services for several weeks. Now that I had my own car, there were a few times that we went out to eat and I drove her back home afterward. I was very happy that I had found another friend - that is, happy until the evening she left church with another guy!

She had not spoken about meeting another man, and we had not really talked about dating other people, so I had just assumed she had chosen me as a male to pursue romantically - but with the understanding (at least in my mind) that we were moving slowly with our courtship because of the age difference. With everything else that was going on in my life, I guess I was so content thinking I had found another romantic possibility that I took Min for granted as a sure thing for those two months.

One Sunday, however, her smile and those eyes that had previously always seemed to be directed toward me were wandering in other directions. Between school, church, and her job, Minnie had lots of options of course. She worked nights at the local movie theater as a cashier - mostly selling popcorn and drinks. Considering that there were hundreds of young men that would flirt with, tease, and ask to take her out on a date, there were more opportunities and temptations for her. As one songwriter wrote, "There are so many flirty guys with pretty smiles and flirty ways." I guess one of their "beautiful" lines had her wanting to find out where that would lead her.

That evening as church was winding down, Minnie left a few minutes early and I had no idea where she had gone. Then, I was left wondering what to say to the rest of our friends from class when others informed me that they saw her leave in a car with another airman. Pain swelled up as the wound from B. J. was re-opened. I was stunned and had no words to describe the internal emotions that came over me. Even though there hadn't been sufficient evidence up to this point in our relationship to account for such a strong reaction, it really got me thinking about how much I cared for and thought of Min.

With my head racing from heartache and confusion, I knew I had to find some way to manage this hurt inside myself. Although I could have discussed the problem or circumstances with the older airman that had brought me into the church, communication did not seem appealing at the moment. Space seemed like the best answer to my dilemma, so I decided to visit other congregations for a couple of weeks. Well, to both of our surprise, apparently my decision to go to Boise those Sundays shook up the church and the Sunday school departments. Because of my abrupt departure, I became the topic of a church meeting during which the realization was reached about how distracting it was having a single man teaching Sunday school to young, impressionable girls. I was clueless and didn't realize that my presence was a problem.

Upon my return to First Baptist, one of the older airmen and their wives spoke with me about how I had abandoned my church responsibilities. They told me that I needed to present an official resignation as a teacher. *"Just come back, sing, and be active regardless of how anyone else chooses to act."* Concerning the Sunday school class, the solution for the problem was a simple fix in finding a woman or older gentleman to teach it instead.

Later on, Minnie told me her mother gave her a tongue lashing over going out with another man when she was known to be seeing or talking to me at church. Her mother was a wise, God-loving, Bible believing woman who told her if she insisted on going out with other men, without a great deal of discretion, that she could be accused of being something that she is not. (I got the idea that something along the lines of harlotry was insinuated.) Minnie specifically told me about the other Tennessee man she dated, but just to point out that she couldn't believe we were both from the same state given how different our accents were. She liked to tease me about the way I said words like fire – "Far, far, far!"

The Noisy Barracks Incident

When listing some of the aircraft maintenance shops or detachments that were inside some of the smaller hangars, the Jet and Reciprocating Engine Shops took up most of the spaces in the main and largest hangar on the Mt. Home Air Force base. Some of the other shops shared spaces, like the Aircraft Electrical and Instrument Shops. My roommate, Tony Costello, was one of the A/1C's who worked with and repaired aircraft instruments. Each of us had picked our room for its location (first floor near the rear exit and the parking spaces) because we both had an auto. Born and raised in New York, Tony was a practical joker and a friend in most cases - if you can call practical jokers friends.

As this story unfolds one might wonder, "With a friend like that, who needs enemies?" We both attended church on Sunday mornings. While Tony was Roman Catholic and went to the base chapel for Mass, I drove to the town of Mountain Home for worship services at First Baptist. I usually stayed in town for the evening meeting of the young people's group and afterwards Minnie and I would go for a drive and/or a bite to eat at one of the fast-food places. My arrival time back to the barracks varied somewhat, but most of the time I would make it in before 9:30 P.M. The 10:30 or 11:00 curfew was more of a courtesy and not a regulation time for us Senior Airmen who resided in barracks # 109 on Flyover Ave.

For the surprise trick Tony was going to play on me, he had to be out of sight but had alerted the airmen occupying the nearest rooms, as in many instances things can go very wrong and this was to be a two-phase joke. First, he had rigged up a bucket of cold water over the inside doorframe. Instead of re-baptizing me as planned, the mechanical part of the rigging failed to work, and the bucket fell at my feet when I opened the door. Surprised nevertheless, I laughed at Tony's attempt to soak me and my clothing. All the neighbors remained quiet in their rooms as I went and retrieved a mop from the utility room and cleaned up the water before starting to remove my civies. The only sounds I heard from the rest of the barracks was soft music playing from some of the rooms as they were all awaiting the next, so-called shoe to drop.

After hanging my clothes in the closet and beginning to climb in the bunk - Wham, Bang, and Thud!! The usually strong, dependable bunk had collapsed! At the sound of the weight of the bed and my 160 pounds hitting the floor, all the lower deck occupants that knew what was going to happen erupted in very loud, roaring laughter. Those unprepared on the second floor, however, (including a Staff Sergeant who most did not know lived upstairs) came down to witness the great spectacle. Loudly, the angry sergeant yelled, *"What in the H--- is going on this time of night? All of you are supposed to be in bed sleeping!"* Then, pushing everyone aside and pointing his finger and fist straight at me commanded, *"Get in the bathroom and start cleaning and don't stop until I tell you that you may!"* I started to explain, hoping someone would speak up in my defense, but most were still laughing and feared they too would receive some punishment. All were incredibly surprised at the joke and at Monroe's overreaction.

Having been first class rank long enough, I knew both Tony and the SSgt were very much out of order. There are military regulations against such dangerous jokes and the sergeant could not reprimand and spot-punish a senior airman without a written report in which the airman has a chance to try and explain his actions. At most, he only could have ordered me to straighten up the bed and room.

I knew my rights and they had been violated. Also, I was on my last year of my 4-year enlistment and, having completed all of my paperwork and diligently working to accomplish all of the shop training for my promotion to sergeant, knew that this incident could be damaging if erroneously reported and placed in my service record. Everyone knew the sergeant would report only his side of the story the next day, so early the next morning (long before the usual time to rise for showers and shave) I quietly dressed and was over at the Command Master Sergeant's office by 6:45 AM. After knocking on the door and being invited in, I said, *"Sir. I have been cleaning bathrooms all night without sleep."* M.Sgt. Wallace asked what happened and I told my side of the story. Wallace then summoned the Charge of Quarters to locate both and inform them to report to his office; saying, *"I want to see Sgt. Monroe and Airman Costello in my office ASAP."* With that, one of my most useless and difficult nights came to an end.

Two Men That Misbehaved and Were Busted

Willis, a cowboy from Wyoming, and Florries who must have been from Texas, were awfully close in relationship. As quick friends, both were jet machinists from my minor repair section. Anytime we saw them in civilian clothes, these two were very popularly dressed and wearing cowboy boots and hats. One of them had a car that they drove around; although it was questionable as to who owned the car at this moment in time. They both were A2/C with the same pay as all A2/C's - about $360 per month. As the story goes, they met a couple girls in Twin Falls, Idaho, which was about 115 miles away down a two-lane highway from Mountain Home. After the TDY to England, these two airmen would be gone almost the entire weekend.

Since we were such a close network, we talked to one another often either at the mess hall or in the café. We presumed that the married airmen took every chance to get home, but most of us could only afford one night out off base each week. These two cowboys would either leave after work, or sometimes would get permission to take off early on weekends, and always return no later than 7:10 AM on Monday mornings. Except one Monday morning they were a "no show" - in military jargon that meant they were AWOL (Absent Without Leave). The Wing Commander had received a call from the County Sheriff requesting that someone be sent to Twin Falls to bail out and retrieve Willis and Florries out of jail. They had been charged with theft. It seems that the reason these two could afford to be gone for the weekend was because they had found a construction company that did not lock up their equipment at night - especially the fuel tanks - and these two cowboys had been helping themselves to the company's gas for weeks, perhaps even two months.

Barracks Life and Part-time Work

An Alaskan Northern had borne down on the base at Mountain Home, and the ground was well covered with snow although the temperature was close to or below freezing. I had not been outside but was instead studying for my GED test (or the Bible to teach Sunday school) and knew I had laundry that must be taken care of. The 9th Field Maintenance was somewhat segregated, with all the shop personnel housed in or near the same street and barracks locations, organized away from hospital and base maintenance personnel. The jet engine section and some of the other maintenances were located in barracks side-by-side with a fire lane (a stretch of open lawn) between them. One barracks had the day room with pool tables and a TV and the other had the utility room with washers and dryers, broom and mop closets, etc.

Taking a break from studies to rest the eyes and relieve my backside, I was probably thinking or dwelling on something I'd been studying when I opened the door and stepped out into what was a snowball fight between one barracks and the other. I heard this loud laughter coming from both sides or ends of the snow-covered fire lane and, before I could put up any defense, I was pelted with snowballs from both sides - extremely hard ones at that. While it was not uncommon, boredom in the barracks often led to arguments over the television or irritation between the airman who wants to read quietly and his roommate listening to extremely loud music, but this was definitely a memorable event worth noting in my letter home.

The Café

My shop supervisor, whom I respected and cared about, had recruited me to work for him in the café there on base. My reason or purpose for working part-time was that the allotment going home to my parents cut into my paycheck and I needed savings for a car - or money for repairs on the one I had purchased and driven to Idaho from Tennessee. Sergeant Krismer was from Texas, and he and the wife and children must have lived in the base housing because his entire family was in the café often. He was one of the top supervisors in the shop and the primary instructor in the minor repair section, as well as a favorite mentor because he knew his Bible. A veteran of WWII, he often spoke of his experiences working on or maintaining the B-29's and B-50's.

One evening while I was working the register, one of his children jumped line and came up to the cash register. When I requested, they go to the proper spot in line because there was a lot of people waiting to order and be checked out, the sergeant displayed a different side than I had seen out of this Christian man before; he gave me a piece of his mind and then corrected the child. Very embarrassed and frustrated about checking out the next few customers, Kris finally did relieve me and sent me to the kitchen where a nice young woman was washing pots and pans with the other dishes. Wanda Jones lived in the town of Mt. Home; she was a hard worker and a super nice person. In addition to being the primary dishwasher at the café, she also cleaned the kitchen, so we had plenty of contact working closely together on clean up late at night. We began joking with each other, smiling and laughing and at times putting hands on each other - nothing improper, just bumping into each other on purpose occasionally.

Wanda was older and heavy built – chubby. Some would have considered her fat, but like most good-looking women who carry the extra weight properly it is not noticeable to those who are well acquainted with them. Wanda had a beautiful face, and it seemed a mystery to me as to why she had no male friends hanging out at the café to take her home or ask her out for a date. There were such few women in the area, especially single and available ones. Well, Wanda had been waiting for the right person to come along but my adult character was just beginning to form; and, after the big disappointment over losing Miss Detroit whom I had someday hoped to marry, my respect for the female gender had changed from super tender to cold, casual respect. I had become a big flirt and was overly friendly with all the ladies I met in Idaho.

Most considered me an extremely nice looking and well-behaved, churched Airman. I was a hard worker, who didn't smoke or drink, used good language, and was laughing and smiling most of the time. Wanda thought she had at last found her man. She no doubt had asked enough questions about me to know I did not have a wife and child back in Tennessee. Well, she showed up in church one Sunday evening to find me and Minnie Dew, whom she knew, sitting close together in the worship service. After the service she asked, *"Why did you not tell me about her?"* I was embarrassed and bewildered that she had taken my gestures in the café that seriously. Yes, I was guilty of ungentlemanly behavior; another lessoned learned by an imperfect young airman – 'Crazy Ray' as I became known 10 years later.

Honored

The date 10/24/56 was probably fourteen months after completing basic training at Lackland AFB in San Antonio, Texas. This is the date I received the letter of achievement for obtaining my GED. This also was the first time that I had been reprimanded since basic. The officer in charge of education programs at Mountain Home was proud to announce that an airman had used the Armed Forces In-State (USAFI) to achieve a high-level education. He had me come up front, up on stage before several hundred airmen. Being embarrassed, I neglected to salute that officer and he dressed me down for it. Of course, he was right and I could only say, "Thank you, Sir" and walk back to my seat red-faced. My intent was on learning as much as possible and doing the best job that I could here.

The Charge of Quarters Problems

Soon after the barracks incident, a couple of other events happened that did not turn out to my liking. We airmen who had been promoted to Airman First Class (E-4) were being monitored for a possible promotion before the time in grade came up because the Air Force, under General LeMay, was looking for those exceptional men to fill the NCO ranks. One such man in the Jet Engine Shop had already been promoted to sergeant; he had tested out in the entrance exams and scored high enough on the aptitude for the officer's ranks but wanted to be in the enlisted ranks. (He was from state of Maine - those people are weird anyway in my opinion.) Sant Allman was a top man on and off the base who was well-liked by everyone and had a strange accent.

I was not thinking about promotions, just trying to do my job at the shop and keep up with expenses. Auto insurance, gas, and minor repairs were exhausting my monthly pay after that allotment to my parents came out of my check. The part-time job that I had taken at the snack bar was helping supplement my income, but also took time and focus. Scheduled to pull Charge of Quarters sometime in the month of November, the base went on another alert, and I was informed that my scheduled C.Q. duty had been moved ahead; although I suspect this was ultimately part of my training in order to check my performance on handling things from the office area instead of the shop environment.

The departing Charge of Quarters briefed me on the log-in-and-out book which contained several names and numbers that I was required to contact during an alert. Well, when I went to work nonchalantly calling those numbers, I was surprised at times by who answered the telephone. (Children many times pick up the phone before the adults can stop whatever is taking place and come to the phone.) There were instances when the babysitter would answer and inform me that the adults were taking in a movie or gone to a party; and one such person said the adults had gone grocery shopping. My problem was, between all the calls and airmen checking in and out needing to be logged and signed, I would become involved with another occurrence or conversation and forget what was said or reported in the phone calls. Later I was going to learn that remembering details was the biggest and most important part of my job.

Finished with the shift, and having been duly and properly relieved, I went to my room in the barracks and promptly fell asleep. After only about 3 hours of sleep, a rude NCO came and woke me up saying, *"Get over to the CQ's office."* Another NCO whom I had never met or seen before began asking questions about the phone calls and my entries (or lack of entries) in the logbook. He explained the seriousness of the non-entries. The calls I was making or supposed to be making were to airmen or sergeants living off base who were supposed to report to their duty post on base promptly when an alert was in progress. In the instance of war, these people were needed to establish, unlock, and open special supply rooms or move machinery and equipment. So, as a result of my failure to contact them or log the calls, these individuals had to be reported AWOL. They could, however, contend that they never received a call even though they were home, in which case the problem came back on me as the Charge of Quarters. I could and might be punished along with those that were informed and then did not respond to the alert. The angry NCO said these men will have to leave their homes and families off base while the men themselves would have to relocate on base. Not through with me, he made a phone call and directed me to another office building where another angry sergeant informed me about the rules and regulations, articles of conduct for failure to comply, and how the punishment thereof may apply to my situation. In the end, I do not know if any of the men had to move back on base, but I did have to go back and redo the logbook properly after getting chewed out by three non-commissioned officers.

Regulations Apply to Everyone

One day Tony brought an aircraft instrument in our room that was quite a timepiece - a combination clock or timer with multiple settings and compass. He had modified the holding device so the instrument could sit on a desk or table. As his roommate, I was extremely impressed with Tony's ingenuity - and the courage to remove such a thing from a plane or the shop supply. Without a doubt, he was one of the top technicians in his section. There were specific regulations against removal, or possession of, any government parts or equipment - especially from aircraft. Airman Costello's shop supervisor must not have been able to explain the instrument's whereabouts to one of the Wing Command NCO's when the shop had inventory and inspection. When Airman Costello was quizzed about the instrument clock, he had to admit that it was in his personal possession.

Regulations had to be followed and Tony was removed from the shop and placed under shop and barracks arrest while the Air Police did a full investigation. Of course, they confiscated the instrument and put it back in the shop. Now that the problem and situation was out in the open and being discussed throughout the 9th Field Maintenance Squadron, the Squadron Company would now be deciding Tony's fate.

Since we were friends, we talked and prayed together; discussing fears over what the outcome would be. He made several trips to the chapel and the Chaplain, Major Dalton, was trying to use his influence in Tony's favor. Tony was an exceptional, good man without any serious misbehavior on his record. Of course many, including his shop super, knew about Tony's practical jokes - especially the one pulled on his roommate. Tony had rights and was instructed to take counsel from the JAG office. Although it was a court martial offense, since he did not take it off base or damage it, they wound up having mercy on him much to his relief.

Aerial Refueling (IFR) and Air to Air (AAR)

If we were to take a ride on one of the military aircraft when they are on a mission, this would be one of the most dangerous rides of your life. We mechanics very seldom fly with the B-47 unless there is a problem that only shows up under extreme conditions.

This story begins with the sounds and screams of the fire trucks and crash equipment going out and taking up their emergency "crash and burn" positions. We were in the Engine Repair Shop, and the supervisors allowed us to try and get a glimpse of what was going on. We learned that two of the aircraft from Mt. Home AFB had been in a midair mishap, but we did not know the extent of damage. The fact that one was coming in on emergency status was better news than that of the aircraft crashing elsewhere. The KC-97 had been on an aerial refueling mission. Although the refueling plane came in and landed without difficulty, a of couple fire trucks followed it while it taxied to the hangar instead of the normal parking place. From where we stood outside the hanger, a damaged propeller on the outboard engine was visible. Parts of the prop seemed to be dangling or hanging loose on the engine. Then too, noticing that the crash crew stayed in position, we knew there must still be another aircraft planning to come in.

The second plane that would be attempting to land, we were told, was a B-47 with a missing vertical

stabilizer. Lots of questions crossed my mind: "How much fuel did he get before they broke off the fuel boom connection?" "How much will the crew try to dump after the alignment with the runway as they talk to the Air Traffic Controllers about the loss of hydraulic fluid?" "Will there be enough fluid to power flaps or slats as they are now called?" They need to stay high enough for the crew to bail out in case the wheels don't want to extend down and lock due to the low fluid. We waited in suspense anticipating the B-47 crew's decision to bail or not to bail out. Only the pilot knows if he can land and try to save the aircraft in this most difficult and unusual emergency.

"They're coming in!" the tower alerts the ground crew. Many of us are praying, some stroke their rabbit's foot, others count their beads, others stroke or flip their lucky coin. *"He made it!! He made it!!"* It still took the entire length of the runway, even after deployment of the brake chute, to stop the severely damaged Bomber after touchdown. That pilot deserved congratulations; seven out of ten would have either bailed out or crashed.

That pilot and navigator were fortunate this time. Without a doubt they used all their training and skill, not to mention intestinal fortitude. Now we mechanics would get our chance to examine the damaged airframe and skin and other places in addition to the stabilizer. The engines will have to be pulled if any limits were exceeded. No doubt the pilot had to use power limits outside the normal to keep control in the turns and maneuvers. This would be lots of work for both aircraft sections.

MSgt. Edward E. Keaton/Bowe (AKA: Buster or Pee-Wee to his high school classmates)

Family and friends knew my "cousin", Ed Keaton/Bowe (or Buster) as PFC Keaton in the Tennessee National Guard at Cookeville, TN. Along with myself, he had gone to annual training with other members of the community from Putnam County. Like me, he served in the Aggressor Force, as it was called, for one training session under Sgt. Arthur Thompson. For some clarity, this was a 'blue against red' simulated warfare game. Because we in Headquarters Company in Cookeville, Tennessee were a small unit, we were picked as the "aggressor force" to attempt to evade a red company of tanks in the field with only some bushes and small trees for cover. The tank company was training with the hatches closed, so they could only see and watch our movements through the range and target finder and window viewers inside the tanks.

We managed to not only evade the tanks but also captured an unsuspecting company of cooks and KP's during the 4-hour exercise. The company and staff couldn't stop laughing at our resourcefulness in that exercise. All the officers and NCOs in the company were expecting us to be griping and complaining about having to get dressed in black and wear a silly helmet with a cone mounted on top. Instead, Edward, Charles, and I were laughing and enjoying our roles - until one of the cooks took exception to the M-1 rifle being pointed in his direction! Sgt. Thompson came to my rescue by telling him, *"This is a game we were ordered to play; sorry you were not informed!"* Once it ended, most saw the humor in how we had made a joke and fun out of the exercise. Edward was a good soldier and a fun guy in the National Guard community.

In 1956, Ed chose to join the U.S. Air Force to see the world - and *that* he indeed did! Edward could have been an officer in the military if he had so desired; he knew something about the enlisted life from the stories of his friends and family members serving at the time he entered. He had graduated from Bixby Seminary with honors as a 16-year-old and attended Tennessee Tech University in Cookeville.

When he was tested at Houston Barracks in Nashville, TN, he ranked in the top 1% out of the recruits in his class and was chosen straight out of Basic to be in the United States Air Force Security Services (USAFSS), which is a tight-knit branch of the Air Force tasked with the monitoring, collecting, and interpreting of electronic signals from countries all over the world.

The USAFSS mission provided the critical intelligence essential to the U.S. in the Cold War effort and beyond. Ed was sent to Yale University's Institute of Far Eastern Languages to learn and become fluent in Korean. In 1956, he was stationed in Korea as a transcriber/translator. During his 21 years of service in the Air Force, Ed was a Cryptanalyst/Radio Traffic Analyst and Surveillance/Warning Controller who was stationed at multiple bases and remote sites around the world, and he received numerous decorations that included the Air Force Commendation Medal.

In his last two years of service, he transferred into the Air Force Recruiting Office. Ed dedicated his life to the defense of our nation and was a mentor to many young airmen as they processed to and from the intelligence community and many other units.

He married a local young lady named Judy Jernigan and they may have had children before the rumored divorce. He will be long remembered in the Bixby community. He retired in 1975 as a Master Sergeant and made San Antonio, Texas, his home. In his retirement, he had the opportunity to travel across the Southern United States.

Life at Mountain Home - 1956

Some time passed and the cold Idaho winter bore down on us at the base. At times it was so cold and windy that we could hardly move around - even in the Alaskan parka that they had issued to me when I first arrived. Although it was winter, work had to be done. When a work order came to the shop, a specialist had to proceed out onto the aircraft line to do the required maintenance on the B-47 engine. We always had to use a construction-type stand to reach and adjust any component on that aircraft engine. We were compelled to triple-check the jack stand wheels, making sure that they were locked, to prevent the high winds from moving both the stand and airman into the sheet metal skin of the aircraft.

To demonstrate the Lordship of First Baptist in Mt. Home, the pastor and staff organized a trip to the State Prison in Boise. We were to preach, teach, sing gospel songs, and give individual testimonies of how God changed our lives. To our surprise, the inmates had much more real and more mature testimonies than we could have imagined! For instance, one inmate cited the benefits of being in prison in relation to his state of mind and acceptance that he committed the crime which brought him to this place. He believed that was all part of God's plan so that his life and soul would end up being saved.

On the way back from the prison, Minnie and I had a frightening experience. The weather had changed and there was now about four inches of snow on the highway. Suddenly my car did an unexpected donut as a semi-truck was approaching. Whatever caused the car to spin out of control is unknown. How the semi driver was able make the necessary adjustments to avoid hitting us is beyond me - it just had to be God's intervention! I did manage to stay on the highway and straighten up, allowing us to make it back to Mountain Home. My Ford was not an automobile made for the terrible Idaho winters. All those miles and the other wear and tear forced me to take it to the repair shops. Several of the older cars stalled out and people were forced to use base transportation again. Airman James Mason thought himself to be an auto mechanic and tried to help me out with repairs.

Sam’s Consequences

Sammy Blain’s character has been described earlier when we were in Detroit. Blain’s exploits in Boise would follow him wherever he would go. At least one of those women who was looking for the same kind of activity he was became pregnant, so her parents called a lawyer, who in turn called the Wing Commander. The nature of the call was to demand some compensation. Sam was summoned before the Wing Commander in order to be criticized; scolded for the situation and the negative perception that it gave the Air Force Base with the public. He was given a choice to either marry the woman or pay the medical bill. He told me later that he knew some of the other airmen had been with her because it had been their recommendations to date her. One of Blain’s stripes was also taken from him because of the situation. Because Sammy chose not to marry her, it is suspected that he was sued for child support which left Sammy just a tad bitter. One of the airmen from the base did end up being a good man by marrying the woman and raising Blain's child as his own. A few months after the child was born, I had a chance to meet this man and found him to be a nice guy. But Blain, still being bitter remarked, *"How are your wife and MY baby getting along?"*

My 1952 Ford didn't hold up as well as I would have liked in the eighteen months that I owned it. It seemed that no matter what James Mason or I did to repair it, something else would just break. Finally, it was evident to both of us that it wouldn't last another winter. This Ford was just not designed for the harshness of the Idaho desert winters, so Mason and I patched it up the best we could to get it ready to be traded in. By this time, I was now a First-Class Airman and could afford more reliable transportation. I was unusually quiet a few miles driving back and forth from town to the base, trying to figure out what I might want to replace the '52 Ford with.

Now 22 years old, I was ready for a serious relationship, but Minnie was still in high school and not ready for such things - most times she just wanted to daydream about the future. In truth, we were both partially afraid on account of what was going on in the world at that time. Some of the other airmen were listening to the news over at the pastor's house with Minnie and I when we learned the Soviets were getting involved with the ongoing revolution in Hungary (from 23 October until 10 November 1956) and our Wing had just been put on alert.

I finally made the choice to trade in my car for a 1954 Mercury. This auto was a beautiful piece of machinery, purred like a kitten even after reaching the higher gears, and it was quick to accelerate and run. I couldn't wait to show it off at the Dew's residence. I hoped that Minnie would like it and would possibly agree to drive it to Boise with me. For our first trip, I took Minnie up into the mountains overlooking a small clearwater lake that had trout fish; which swam in schools toward us as we stood out on a fallen tree trunk. I thought to myself how rarely I'd seen any fish swimming this way when I was still living at home. In Tennessee, we had what we would call 'the blue hole' where we would swim around, but for the best fishing we would have to perch ourselves up on the bank out of sight, careful of not making a shadow. Later in the evening, when Minnie and I drove back down into Mountain Home valley, we happened by some small streams where we would occasionally spot a beaver. (Beavers don't linger out of the water very long.)

The Rodeo

A few weeks went by before signs and banners about the coming rodeo started to be advertised in the area. It was called 'The Snake River Stampede' and was to be held on a given date in Boise, Idaho. That's 50 miles away and I wasn't too excited about driving over there by myself, although seeing a real live rodeo would be another experience. Min and I talked about the rodeo and how much we wanted to see the "mutton busters" - the rodeo kids trying to ride a ram or ewe. We also dreamed of watching the bull riding as well as the steer riding and roping contest.

We finally took the step to talk to her parents about making the trip over to Boise to go to it. We were somewhat surprised when they agreed and allowed us to go because we would have to be out until very late. I learned that I was in the trust zone with Mr. & Mrs. Dew and that the rodeo was both a treat and a test for me concerning their daughter. The Dew family was to be commended for the way they adopted this young man far from home who was in constant need of some home cooked meals - and the 'mom and dad' figures. The welcome mats were constantly out. Any time my car was parked beside the curb at their residence, I knew that was a real home - unlike the one that my Uncle James Bowe Jr. described as us being "Jerked up in." My parents loved us, my brothers and sisters, but none in my childhood home had ever known the real family unity consisting of love and affection that I saw in these far western Americans. Was I special? Yes, very much so! I was made to feel special by the Dew family because I was one of the church members - but having their daughter interested in me didn't hurt either.

The church and family were of great importance to Mrs. Dew, and I was slowly realizing the implications of how my actions while in the trust zone would reflect upon her heart in both of those categories. Although we knew that with their trust came responsibilities, both Minnie and I were ecstatic with joy at being able to do something as big as going 50 miles to the Snake River Stampede together. We had a fantastic time together climbing up on the fence and watching, with intense excitement, the dangerous spectacle of the rodeo! Even the hot dogs made a lasting impression as we watched these cowboys get bucked off those wild horses and while riding the bulls - even though it looked so dangerous. (I could never be able to imagine anyone putting their health on the line like that.)

The rodeo didn't get over until about one o'clock in the morning. I decided to let Minnie drive the Mercury while I slept in the passenger seat. Suddenly, I was jolted awake by swerving and a quick application of the brakes. Seems that she had barely missed a cow out in the middle of the road. That area in the Idaho wilderness was still an open range for cattle. While the herd is usually kept far away from the roadways to avoid situations such as this one, there are always a few stragglers.

I wished that I could have slept the next day, but there was work to be done and there was word going around the base that another T.D.Y. was coming up. Usually I'm the first to be excited to go travel, since that was a big part of the reason why I joined up in the first place, but this time was different and unusual for myself. This time I was thinking about how much I would miss Minnie more than the excitement of going wherever the Air Force wanted to send me.

Airman B. H. Bowe's Orders to Donaldson AFB, S.C.

Meanwhile in 1957, my brother Bud H. Bowe's job title during his second enlistment was that of a Load Master: someone who secures the load, balances it, etcetera, in the cargo area. The Air Force does manipulate the good men they want to keep. Once you receive your "job" training, you typically serve out an apprenticeship for three years until you achieve the rank of Airman First Class; this is their way of filling those rough, not so interesting jobs that must be done for the successful completion of the Air Force's mission and operations.

We did not know whether to call his orders to the 63rd Military Airlift Wing at Donelson AFB in South Carolina a break or a calculated attempt by the Air Force to try to keep a partially trained man on flying stats. In my opinion, the cross-training from a Pole Handler on the C-119 to a Loadmaster in training on the C-124 favored the Air Force rather than my brother.

The Air Force had the very minimum invested in A/2C Bud H. Bowe at this point in his career. Versus having gone to one of the Air Force schools, his was an OJT (On the Job Training) position - considered low skill but an especially important job. He should have been promoted to an A/1C a year before this time in his career. After enduring all that, the Air Force must have taken notice that they had a valuable man in their ranks.

A Tale of Three Brothers

No two individuals are the same. Although brothers, our character develops within the family as we grow and interact with each other. My view of him as his older brother was, and is, that Bud is one of the nicest men you would ever want to meet. However, being nice in the Air Force does not count for much when you first go in.

One part of his character that kept him "on the back row" so to speak, in my opinion, was that he allowed life to come his way instead of asking questions and challenging situations that he faced. For example, on the farm, Bud had three bosses: Dad, Mom, and his older brother - me. While he could ignore some of my orders and work around some of Mom's, Dad was 'God' and highly abusive to the family. Only J. C., our youngest brother, ever challenged or questioned Dad. (I guess that's why he felt it okay to constantly challenge his older brothers when it came to doing chores and work in the field.)

An example of the family situation (who was privileged and who was the underdog in the family) was when us three young men were ready for high school - well high school for me and junior high for the youngest. Being a poor sharecropper, W.W. Bowe was concerned about feeding his family of eight. Dad had taken us boys out of school whenever he needed the extra help to accomplish a particular job. At this time, he was planning for the next year's crop, after the harvest and tobacco sales of the past years had not done very well and his other job (clearing the timber off the Center Hill Reservoir) was completed. So, being the only breadwinner of the family, Dad alerted us that he was definitely going to need an extra hand in the field the coming year.

Two men were required to accomplish the many tasks of working the mules, turning the soil, planting the seed, pulling treetops for firewood, etcetera. As the oldest, I had already spent several seasons away from school for this reason and didn't welcome another interruption - surely the other two felt the same way.

Whether Bud volunteered or was drafted, I am not sure. I do not remember protesting or trying to make a case against staying out of school an entire year, but Mom may have been advocating or lobbying for me behind the scenes. As her firstborn son, Mom favored me, and I was highly privileged in many ways due to this. [Later, I recall thinking about this matter and knowing I owed my brother a year of service.] Not only did J. C. and my father not get along, he also didn't have as much experience with the team of mules that we did. Buddy was more of Dad's favorite. The point of this story is that Bud H. Bowe went into the service with only a 9th grade education. His experience was limited to farm work, so being selected for a top-secret job that did not require math, science, reading skill or any special education was a blessing.

I did not how my youngest brother, Jess, rated and was selected for Air Frames or Aircraft School in Wichita Falls, Texas; one of the largest Air Force schools in the country. Somehow or another, he had managed to enter the Air Force though he was actually only sixteen! Bud and I were in wonderment; how did he get in to begin with and selected for a regular school and such a historical job? Turns out that he was not truthful on his application and was thought or assumed by the Air Force to be nineteen and a twin to Bud.

J.C. had also enlisted in the Air Force (in December of 1954) and did his basic training in the state of New York. After basic his orders took him to Shepard AFB for Aircraft Mechanic School. Upon completion of that, he was assigned to the largest aircraft the Air Force was flying (the ten engine B-36) and stationed at Biggs AFB near EL Paso, TX.

Now in his last months of the four-year enlistment, the Air Force has transferred A/2C Buddy H. Bowe to Donaldson AFB in Greenville, South Carolina, where he will start the training to become qualified as a Loadmaster on the C-124 Globemaster II - the aircraft nicknamed "Old Shakey." Certain qualifications must be met Age limit between 17-39, must have normal depth perception, height between 67" and 77", must pass a national and local security agency checklist, and pass a Vocational Tech battery of tests. (These were waved for prior aircrew with qualifications up to date.) The Loadmaster's were responsible for supervising the loading and unloading of cargo, vehicles, and people on a variety of aircraft. This involved mathematically planning the correct placement of the load on the aircraft and included working closely with flight engineers to fill out the weight and balance sheet that is passed on to the Pilot and Aircraft Commander.

"Sputnik" is Here

One of Khrushchev's grander schemes was the building of a new city for science and technology in Akademgorodok, Siberia. One young scientist said, *"We wanted to catch up with the West."* And catch up they did, at least in some fields, such as space exploration. The Soviets were pushing ahead and in 1957 they launched the first satellite. 'Sputnik' orbited the earth in 96 minutes, doing 1500 orbits in all. This feat, which elated the Soviets, frightened the leaders of America and Western Europe because they were concerned that the Communists could win the Cold War if they weaponized their satellites. The western leaders tended to forget that most of the Soviet population could only watch the satellite streaking through the sky from their place in the bread line or the collective farms where the derelicts were sent to work. It was an awesome sight to watch the extremely small, white ball crossing the sky. We watched with amusement and wondered just what it meant for us airmen on active duty.

Bud's Second Crash in Cambridge Bay, Canada

In April of 1957, during Buddy's second enlistment, he and the crew on the C-124 (also known as the Globemaster II) were making an important delivery of a fire truck, vital generators, and other overdue equipment to one of the 57 manned stations on the Dew Line (Distant Early Warning Line) near Cambridge Bay in Canada.

The pilot misread or misjudged something and landed the aircraft short of the cleared area. This truly was a short runway and the C-124 was heavily loaded. We would have to read the pilot's report to know what really happened, but the results of the crash were easy to observe. Somehow, perhaps by hitting some directional landing posts, both of the wings had been stripped off the aircraft.

"When the aircraft came to a complete stop" Bud said, *"I opened the bubble hatch at the very top of the aircraft and I saw some fire!"* Panicking, I opened the escape hatch, threw the escape rope down the side of the aircraft and slid down the 18 feet; burning the skin off my hands in the process. The instant my boots hit the snow-covered ground, I ran as fast as my Tennessee legs would take me away from the aircraft.

When the rest of the crew came out of the C-124 they were laughing and talking to each other about him. Because there was no damage forward of the stripped wings and the crash crew had extinguished the fire, the others had been able to safely exit the aircraft normally.

At last, the station had received the long-awaited fire truck and equipment that was critical for their operation. They had not, however, anticipated it arriving in such an exciting fashion! Only one crewman sustained some undisclosed injuries.

To get a larger picture of the C-124 Globemaster II, nick-named "Old Shakey": It has a large cigar-shaped body, 2½ stories-high off the ground, long wings set low, and four large Pratt & Whitney R 4360 piston engines producing 3,800hp each. The C-124's featured two large clamshell doors and a hydraulically-actuated ramp in the nose, as well as a cargo elevator under the aft fuselage. This aircraft could carry 68,500 lbs. of cargo to include tanks, guns, trucks, and other heavy equipment. While in the passenger-carrying role, it could carry 200 fully equipped troops on its double decks or 127 patients on stretchers and their attendants. The cockpit was up front on the top level, sixteen to twenty feet from the nose. The flight engineer panel was behind the co-pilot's seat and there were three bunk spaces behind that for the crew to rest since they carried two crews on all long flights.

Sept. 1957 TDY to Guam

When the Ninth Bombardment Wing received orders for the deployment to Anderson AFB in Guam, most were excited because this was the island paradise we had been looking forward to from the outset of our time in the Idaho Desert. Some of the air crew welcomed the move out across the Pacific with the island hopping, aero refueling, and new scenery from air and on land. Others were unhappy about leaving their families behind for an extended time; but so it goes in the Air Force. Most of the B-47 air crew are on their second enlistment, or they are career officers who have accepted the risk of the job for God and country and perhaps finances, and other reasons known only to themselves and maybe their families.

3 Aircraft formation of B-47Es of the

We, as maintenance people, were not informed of the deployment details. We never knew where the entire wing was going or if only parts of it were going. It was almost deafening and, with all the smoke from the JP-4 and fumes from the jet-assisted takeoff fuels, the whole base was cloudy and noisy for the next 3-5 hours. We did know that, once the alert was given to the commanders, the 'War Plan' would be activated: the aircraft would be loaded with a specified 'flyaway kit' and flown overseas to carry out strike missions from the bases there. That was the plan that General Lemay had for the Strategic Air Command.

We in the Minor Repair Shop didn't even know for sure that *we* were on an actual alert; there were so many and so often. When SSgt Hauff informed us that a 90-day TDY was going to be at Anderson AFB, Guam, with some time on Waikiki Beach, we were very excited that we were going to have a layover on the island paradise of Hawaii. Our NCO was excited that the dull routine of reporting to the shop each morning and dealing with unhappy airmen who were merely serving for a paycheck (instead of the nice career that the Air Force offers) would be broken up.

Now our superiors were going to find out about the quality of the new airmen (those that had missed the deployment to England) and those of us who were detailed out to security and duties other than engine repair - like test run-ups out on the test stand or in the aircraft for adjustments. There will be an enormous workload when we reach Anderson. All 45 aircraft were going to require some adjustments and/or engine changes due to exceeding the manufacturer's time limits and other recommendations. (The manufacturer establishes the recommended engine hours that are safe. There is a safety time limit on most aircraft components - especially the engines.)

We in the lower ranks are viewed like apprentices, practicing and perfecting our skills and knowledge of the jet engine components and their functions. We are also semi-packed and ready to travel anywhere in the world at any time, except for our personal items. In the field maintenance departments, we have more time than the air crew maintenance. At times, some practice the routine of checking out the items required for the particular job assignment from supply rooms or departments because in each deployment there are last minute or catchup repair orders that must be taken care of. (The NCO in charge usually assigns the senior airman or new sergeant the job.)

Finally, the formation rollcall for last minute instructions and questions about who loads the personal and shop gear is sounded. (Some small items needed to be handled in a delicate manner.) We must remember that, once on their aircraft, the Navy is in charge - not the shop NCO but the Navy personnel regardless of rank.

The bus will take us and our personal gear out to the Navy R5D which is the Air Force version of the C-54 Spymaster Transport. We had already been introduced to the seating arrangements; the Navy turned the seats around facing the aft part of the aircraft in 1955 when they transported the field maintenance group to England. It is now late September 1957. I wrote earlier about how we always enjoyed the stewardess demonstrating the May-West. We applauded loudly and she smiled graciously. The Navy aircraft had originated from NAS San Diego, CA or NAS Whidbey Island, WA. The flight schedule is to land at Morfitt Field in Mountain View, CA. There, we will be able to see the largest aircraft hangar in the world which was constructed to house the dirigibles - like the Hindenburg that burned. This hangar is so large that a stunt pilot flew completely through it. It is such a treat to see the man-made wonders of the world and it is one of the reasons I enlisted in the Air Force rather than one of the other branches of the military.

------- Sidebar ------

The Boeing B-52 Stratofortress was the first strategic jet bomber with the ability to strike at targets anywhere in the world. This capability was demonstrated by several long-range flights made by the aircraft. In November of 1956, eight B-52b's from the 93rd Heavy Bombardment Wing at Castle Air Force Base, CA made a 17,000-mile flight around North America and over the North Pole, staying continuously airborne for 32 hours.

------- End Sidebar ------

The Navy will refuel this R5D, and we will be off to Hawaii. Even though it has been 13 years since the end of WWII, I was still anxious to see if any of the damage resulting from the Japanese attack back in December of 1941 was still visible. What a long, boring flight! Guess I can catch up on my Scripture reading. There will be some books passed around, along with a loud card game going on, because some of us are too excited for sleep. All there is to see out the windows is water and more water. Occasionally there will be a ship sighted and the commander will announce it, as well as the side of the aircraft it may be viewed from and the approximate location for the passengers to be looking. (Large ships seem very small from 26,000 feet.)

Before the first sighting of the Hawaiian Islands, the pilot will turn on the radio and tune in to one of those easy-listening Hawaiian music stations. That is a wake-up sound to those that were sleeping and a very welcome sound to those who are anxious to put their feet on the soil. They announced that the plane would circle so that we may see the Honolulu Bowl before landing but that also means we are going to come down fast which will hurt our ears.

Now we all get to stretch our legs and walk around in the terminal. Oh those slot machines are just waiting to take your money - if you have any. When we deplaned, the Hawaiian dancers came out and placed a welcome wreath around some of the older Sergeant's necks. We were instructed to remove all our personal gear from the plane so a bus could take us to a military transit barracks for the night and told we will be free until the next afternoon. (We have passed through a couple time zones and gained a few hours.)

Yes, there is still some evidence remaining that the war had damaged the Islands: some partly submerged ships, some old rusty boats, and a submarine lying on its side. I noticed some buildings with bullet indentations that had not been plastered over; it was as if the property owners deliberately left the scars of war for tourists or as a reminder of what took place and could happen again. Naturally we had to go down to the beach to see the bathing beauties and go for a quick swim. Couldn't leave beautiful Hawaii without a trip to the beach!

Fun and relaxing time is over so soon; we hardly felt we had really seen or done anything. Fun time passes so fast, and the work activities pass so slowly. Schofield Barracks wasn't what I expected; it was a huge place with new buildings - no resemblance to the movie "From Here to Eternity" with that star-filled cast of Burt Lancaster, Deborah Kerr, Ernest Borgnine, Frank Sinatra, Donna Reed, and Montgomery Cliff. It is difficult to say where that movie was actually filmed.

Look where that bus just took us; that is a Super Constellation Navy Aircraft! We are going to fly in style out to Wake Island in the most modern flying machine in the world - some more classic air transportation. That means a great deal to this airman who is still a bit nervous about flying. We departed Honolulu in late evening for a fuel stop on Wake Island, then the final leg of this trip to Anderson AFB, Guam. Everyone is tired and we will attempt to sleep as much of this flight as possible. There is nothing to see but ocean, the stewardess, each other, and these ugly but very useful Staff Sergeants. We respect them very much even if they are sometimes nuisances.

Most of us on this flight will be promoted to Staff Sergeant E-5 within the next 18 months if we stick with the program we enlisted for. Three of the airmen we started training with in the 9th Field Maintaintence Squadron have already made that promotion. We are now touching down on Wake. It is always a nice sound to hear those tires squawk when they touch the runway. We taxi toward the terminal and the off ramps being rolled to the doors that will be opened. *"Everyone off for a leg stretch as our transportation is being refueled."* The NCO in charge said, *"Don't get carried away sight-seeing; we will only be here for 45 minutes."*

From where we deplaned, there were only a few palm trees and other aircraft to see. There are all types of aircraft here that stop over for minor repair and refueling: fighters, bombers, sea planes and commercial aircraft. Shortly, we can see the portable ramp being moved into position and know it is time to board. We are loaded and off again for another boring 7 hours if all goes as planned. Everyone is still tired and not in the best of moods with no dancing girls or beaches to visit on Wake Island at this time. Our destination, Anderson AFB, Guam, is finally the next stop.

GUAM

The Island of Guam is a great place; it is closer to Asia which seems to be the trouble spots. The sweet sound of those tires squawking on the runway pavement says we have arrived and reached our destination without having lost a man. It was not so back in the year 1944 when many men were lost taking this island back from the Japanese. But this is still the Cold War, and we are very aware this place is a hot target that would potentially be one of the first sites a nuke would strike.

Guam is the tip of the spear for the defense of the United States. The B-50's in WWII flew from this island and saturated Japan with firebombs before the atomic bomb was dropped from a flight out of another SAC Base Island. We must dismiss that thought and get busy with unloading and finding a place to sleep after that long flight. The GI bunks in the transit barrack are great when the body is so tired; we could hardly stand after the 15 long flying hours total it took to get to this beautiful island paradise we had hoped for.

Upon arrival the first day and checking into the barracks, a set of football pads was hanging on one of the bed posts of a bunk. This scene put many questions in my mind. Some of the questions were: What is this football equipment doing in the barracks of the 9th Bomb Wing? Have we been directed to the wrong barracks? Why isn't this equipment in a locker? - It smells and appears molded. Very tired from that 8-hour flight, I was too worn out to investigate further; the bed was inviting, and I quickly made use of it. Being excited over the trip and new scenery occupied most of us so we hadn't talked much about sports in my circle of friends on the aircraft coming over. Airman Greer from Florida had talked about his time as a high school quarterback. I had not seen him in the chow hall, or on any flight, or in the transit barracks we occupied coming over. [A note about the transit barracks: They are WWII type, open-bay buildings with double-decker bunk beds that the lower enlisted personnel use for one or more nights during travel from one base or island to the next. E-5 and above had more favorable accommodations. Non-Commissioned Officer's (NCO's) ranks start at E-5 and go through E-9.]

A couple of days passed before the questions about the football pads were answered, when one of our own mechs was identified as a football player. Each branch of the Armed Services has a football team; they play all year round as a league, having a set schedule with trophies for best teams and performers. One of the 9th squadrons had an airman who played football at University of Virginia Tech and the military had worked out programs with these college athletes to play as entertainment for the troops. I viewed a newspaper publication with a sports section listing the standings and some stats of the team's stars. Speaking of entertaining us, Elvis Presley was a draftee serving as a Private in the U.S. Army at the same time I was serving our country. Later on, there would be a USO show coming to the island while the 9th Bomb Wing was there in Cold War defensive-mode some 90 days.

Although some of the new jet mechanics did get put in other places, management didn't assign me to other duties as before so perhaps I have gained some status and trust after these two plus years. Assigned to the J-47 Jet Engine Shop, we had loads of work orders for minor repair and engine changes. I had written before about those forty-plus bombers flying nonstop directly to Guam. They were aero refueled somewhere over the Pacific on the way to Anderson, AFB. Pinky and I worked in the shop on the 'out of limits' broken, overheated engines for a of couple weeks before Tech Sergeant Morgan wanted me to work with his crew (The Engine Change crew). I worked with them two weeks, then requested a return back to the Minor Repair Section for personal reasons and spent two weeks as crew leader, then as supervisor over three new Airmen First Class.

During my time there, the Guamanians and people from the surrounding islands had a festival celebration highlighting the native dances, art, and foods of the region. They brought many of their special dishes, along with rare fruits and drinks, strange-appearing fish, and beautiful decorations. All were welcomed and invited including the military personnel. The local homes opened up to everyone as well, but I wasn't brave enough to venture into the native houses. There was an incident that occurred when one of the airmen from the 9th drank too much of the wine and decided that one of the native's wives was his girlfriend, resulting in a severe beating and time in the island jail.

Other activities for me included going to the swimming pool each day and down to the hobby shop with James Mason who was tooling his name on a belt. As he worked with the leather, selecting and using the different tools that would form the correct pattern of impressions, it was truly a work of art. I tried to copy his work but soon realized that the time to learn the skills to successfully complete any project was going to require several sessions and I did not have the patience for that.

R&R Air Force 7 at Kokoda AFB, Japan

Having a 35mm camera and a few days change of clothes always packed, when I was told I had been selected for a trip to Japan I was excited! I do not know how my name was chosen, nor were any other people from our shop or crew on this trip, but my guess was that the retention officer probably added my name to the manifest in hopes of convincing me to renew my enlistment. At some point in the past, I could have written where I would prefer to visit if a flight became available, and I would have chosen: #1 the Philippines, #2 Singapore, and #3 Japan. Despite being 13 years after WWII, there was still some evidence of the terrible war visible in these countries.

The time was probably mid-November of 1957 as we loaded onto the bus that took us to the awaiting aircraft. One of our KC-97's had been stripped of it refueling duties and semi-fitted for passengers; bucket seats had been added in addition to the side-wall seats on each side of the cabin's cargo hold for the flight. The aircraft was part of the 15th Air Force and the 9th Bomb Wing from Mountain Home, Idaho.

ttps://donmooreswartales.files.wordpress.com/2011/06/donald-gatrell-b-47.jpg)
3-47 strategic six-engine jet bomber climbs into to the sky in a hurry with rocket-powered assist pods attached to its fuselage. Air Force picture provided

As usual, I prayed on takeoff as we left the runway at Anderson, AFB in Guam to fly those 9 or 10 hours to Japan. For a time, I wasn't sure where we were going. Some rumors suggested that an alert might go down, in which case we would have to return to our duty stations. The B-47's would be loaded with fuel before they headed out over the ocean to their targets. There were 67 of us, plus extra crew, which went on the R&R (Rest and Relaxation) trip. [In order to satisfy Air Force SOP (Standard Operating Procedure) requirements, the captain or co-pilot have to record the purpose on the flight sheet.]

The flights are rough and boring. Those that can sleep do so, others play cards or write letters, and some told tall tales about the war, or their dad having been in the Army Air Corps as flight crew on the various bombing raids. Some of the older airmen were sergeants, having been in the service that long, but I would say about fifty of us were kids under 25 years of age.

Japan is an island the size of California but the climate here was different; the temperature for one thing was in the low 40's or even high 30's and a light snow covered the tarmac. Who had thought to bring a coat? We sure hadn't needed jackets back in Guam where it had been 80 degrees when we left. My first purchase here would be a new jacket, but that must wait until after an orientation about the dos and don'ts while we are here.

Strangely enough there was no one on this flight from my shop or section that I knew or had met before; only these two airmen that seemed to have the same interests as myself. We sat together during the briefing before we departed the ramp area and were paired together by one of the NCO self-proclaimed tour guides. We were told to stay in a group and that some areas of the city were off-limits to Americans (and especially to the military) because some of the Japanese have not accepted defeat and hate us. They gave us directions and suggested places to purchase items. Additional warnings about the social diseases and spots to avoid because of pickpockets accompanied the stern advice, *"Do not go anywhere alone; make sure there are two, preferably three or four, in a group.*" They told us that 98% were great, honest people and the crime rate was exceptionally low.

What a sight to remember! Most bases appear the same since American contractors built them, however there were jet fighters all over the place and some of the guys started taking pictures of those interceptor planes. At the first sight of a flash bulb, the Air Police showed up and confiscated some cameras. They promptly unloaded the film from them and returned all but one - there must have been something different about that specific camera. He explained to us that we were in a restricted area and gave us orders to move. The new fighters on that base there on Kokoda must have been 101FB's (capable of carrying the nuclear payload to short distance targets) but that is just a guess because of reading about them. Although this was during the Cold War, the manufacturer put out enough info on new weapons so that the enemy knew how to identify and defend against them.

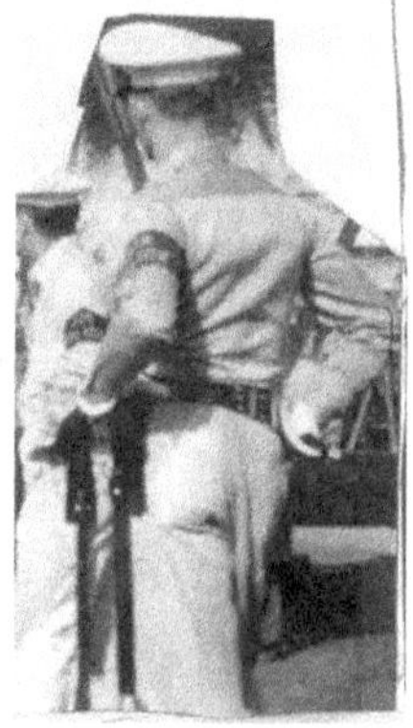

My new friends and I went walking, wandering through the streets and alleys of Kokoda like lost kids, to get a view and understanding of how the Japanese lived. As we scouted out the city, we saw something unusual in the marketplace: the largest carrots, turnips, onions, beets, radishes, and potatoes I had ever seen were among other vegetables that I could not identify or describe. So before touring the fish market, we found a Japanese native that could manage some broken English and began to ask questions about the oversized produce. Then, an interaction with a fellow American solved the mystery. The man showed us their toilets and explained that the natives used human waste to fertilize their small gardens in order to stretch every inch of such a small place to use for growing something edible.

Next, we visited the fish market with the small squid, swordfish, shark, salmon, and other saltwater fish. After that, we were getting hungry but, since we obviously didn't speak or read the language, we were somewhat concerned about how difficult it would be. Well, it was easier than we could imagine.

The Japanese are very clever and know how to sell. In the front window of the restaurants, they placed a plate of food upon a tilted board of bamboo so the foreigners could not only see the selections available, but the price in yen as well. We had to do some calculations to determine the price in U.S currency we had exchanged earlier. Once seated, we guided the waitress over and pointed to the dish we preferred.

When we paid the bill for our food, we gave the cashier too much money and let her make change so we could give the waitress a nice tip. As we continued to walk the neighborhood, I noticed a dug-out trench in the yard of a small business or home which, upon closer examination, appeared to be an angled ditch deep and wide enough for a gun emplacement - like a foxhole that had not been filled in. The grass had grown up around the opening and was neatly trimmed as the natural lawn.

The trains and taxicabs were a memorable sight to see as they came through the city streets and rails with whistles and horns blowing! People there use the streets and tracks for walkways as well as the very narrow sidewalks. Taxicabs approached without slowing down one bit and the shoppers or workers on the crowded streets just scatter like a flock of birds. Smaller than the trains in the United States (and much less safe for the passengers it would appear) the seats were always loaded to full capacity, including many people hanging onto rails; their brightly colored packages and other items not only stacked and tied on top, but also strapped on the sides as well. It is a mystery that many people survive the day, going to and from work, shopping, or whatever things they leave home for, as there is seemingly no speed limit in that part of the city. There must be some fatal accidents with both the trains and the taxis going at such speeds. Those small, three-wheeled bicycle trucks are quite a conversation piece too. They hire out to make deliveries from businesses, hauling their products to market in the back like pickup trucks in the States. Another thing of note: Japanese people are smaller in general than their American counterparts.

Our next move after eating and window shopping was looking at the buildings; both the old ones that had never been fully repaired and the new ones built with American materials but constructed with characteristic Japanese architecture. As it was getting later in the evening, the guys wanted to shop for a hotel, motel or house that takes in guests when they were approached by this one Japanese man and offered a place for three or more. Even though I was very curious as to the events that had so many airmen talking and laughing so much, and I was on this trip for the enjoyment of any once-in-a-lifetime, worldly entertainment that could be offered (some of which may not be legal in the USA), I had my reservations about what the activity might entail and knew I might not handle it well. Although I was resistant to the idea, the two Airmen with me were already primed for a Geisha house so I agreed. We men sometimes do these unspeakable things and hope that the ones that care about us never find out, but eventually they do.

For whatever reason, the program must have required a minimum of at least three paying customers. There may have been more girls that were sent home or to other houses for the night when the numbers were

low. Our evening included the meal that was prepared and served for us, and our companions the entertainers, and also the services of the 'Papa-san' who managed the Nikkei show and kept order in case the GI's drink too much and get rutty. Since my friends could not be accepted into the house without a third person that would pay and stay for the meal and entertainment, that left them having to partially pay for me since I wasn't prepared for this type of expense. When the man collected the money from each one of us, I didn't know or understand the cost because it was in Japanese Yen. We were to stay all night and be served breakfast the next morning.

Following the meal, the girls put on quite a show in Japanese, played American music, and then had a radio program on in broken English. My paying friends chose the two most attractive girls to sit near after dinner which left the remaining person and me together. This lady was much older (perhaps thirty or more) and not as attractive as the others, but nice and seemed to be the more experienced entertainer. She was so much fun because she knew and understood American jokes and even told me some stories about how her people had to live during the war. Some finger food remained on the table, so we snacked and laughed it up until the morning. I wonder who wore out first, her or me. The two of us were so busy I had not thought to check on my friends before turning in.

We all slept in late then had a traditional Japanese breakfast. Since we GI's usually have eggs, toast and a meat, and I did not recognize most of the items they served, I was hungry and tried not to reject this unknown fish and rice meal prepared so well. In comparison the next morning, the other girls seemed dull and tired and bored with us soldiers, while my lady was still going strong and talking about how happy I made her feel. Joyous and hungry, the two of us ate and laughed and ate and laughed some more. I had plenty of tea while she and the others drank Saki.

My lady friend still appeared ready to go all day and another night, but we had a plane to catch. 'Papasan' politely ushered us out of the place. Although I suppose it was an act and part of their business, some of the girls were crying and did not want to turn loose from the farewell embraces. We had been perfect gentlemen, treating them with utmost respect. (It would be difficult to know what these women had gone through to bring them to where they were at this time in 1957.) Next, we had to make connections to get back on the base and locate the rest of the group that we flew in with who would be looking for the flight back to Guam also. There were more KC-97's on the tarmac besides the special one that had been prepared to fly us passengers on this R&R trip.

Brief History of Japan and America

Neither government wanted a great deal of fraternizing between the GI's and the starving Japanese women. At first, based on the assumption that American soldiers would "use and abuse" the women like they themselves had done when they invaded China, the Japanese sent their young women and girls to what they considered a safety zone. By the time we visited, Japan had gone through 10 years of training and education to prepare them for Democracy. The British and Russian soldiers had forces there as well as the Americans.

President Truman wanted to make a peace treaty with them, but first he insisted that the Japanese government agree to a bilateral security treaty wherein America was their only ally; it also gave Washington the right to bases in Japan entirely outside the local government's purview. U.S. forces said the treaty would contribute to the security of Japan against armed attack from without and include assistance to put down large-scale internal riots and disturbances within Japan at the express request of the Japanese government.

Yoshika also had to declare that they would not enter into any agreements with the Chinese Communist government - only then could the peace treaty be signed. Predictably, the Soviets refused to sign it and China was not invited to the meeting. Over time, Japan would develop into the most important U.S. ally for fighting the Cold War as well as the Korean hot war.

Departing Japan

Our flight from Japan back to Guam was filled with surprises. We must have been an amusing sight to behold as the 67 passengers and the KC-97 aircrew made our way to the aircraft parked on the tarmac of Dakoda AFB carrying all those packages wrapped up in soft, colored paper of all colors: Blue, gold, yellow, white, purple, red, turquois; and the mixture of three or four colors on some of the bright, beautiful ribbons. Most airmen had purchased a gift for wives, sweethearts, and family in hopes of later delivering them home to the states. Some special gifts could not be wrapped because of their irregular size and shapes but the crafty Japanese still managed to decorate them skillfully and beautifully.

Just like the scene when we departed the Geisha house, some Airmen had women at the entrance gate crying and pulling at their arms and sleeves for the men to stay or take them onboard the aircraft. I am sure part of the tearful episode was an act or part of Japanese culture; the women's way of saying thanks, come back soon. Instead of purchasing gifts, we three who explored and toured Japan together only took photos and spent our time sight-seeing, exploring the countryside and marketplaces through various modes of transportation. There was so much to see and visit, so much excitement for us to enjoy riding in those rickshaws pulled by strong men. We also rode in the motorcycle sidecars with a Japanese operator.

The ground crew is moving the portable ramp to the aircraft for loading. This is going to be a slow process as the loadmaster decides where some of the gifts and items can be stored and secured for safe flight. His job becomes complex in these non-military, more relaxed type operations. Each man wants to see where his gift item will be lain in the storage space to eliminate an awkward situation.

Then finally, once we are all aboard and accounted for, we were told to take our seats and strap in for that long ride to our temporary home away from home. We were scheduled to make Anderson Air Force Base by nightfall (or 1900 hours Guam time) as we were going to lose 2 hours in crossing the time zones.

The aircraft begins to shake and vibrate as the reciprocating engines warm up while the crew are carefully checking engine and flight controls. We can only sit and wait for the takeoff run as the loadmaster checks all for fastened seatbelts. The aircraft commander announces *"Set Take-off Condition"* over the intercom, and it's repeated by the loadmaster. Those words are like music since we have been sitting for what seems like an hour from when fasten our seatbelts had been announced. It could be that all was not well in the flight station. All of us were anxious to get going; we have had our good time and want to get back.

The take-off run and climb-up seemed normal; the aircraft is flying smooth. When cruise altitude was reached, the Japanese air-controller designated the flight level. Once the air-conditioner and pressurization systems became noticeable, we could unbuckle and began moving about to establish enough players for card games and other ways to amuse ourselves. Some passengers began a conversation about the culture, the marketplaces or city shops where they bought their gifts, or about the restaurants and problems ordering the meals. The largest buzz was about the Japanese women and girls; how nice, gentle, and apparently grateful for any smile and friendly gesture toward them. The "businesswomen" had such an unforgettable welcoming smile; they treated all men, including their own, like kings.

Although the prostitutes were trained in the art of the comfort and pleasure of men, I don't know how they sacrificed themselves to be a servant for males to use and treat as they wished. Some of the married men, sad to say, were attracted to them. The young airmen wanted to find that special, more innocent young woman. Some of the more experienced airmen, like the loadmaster on this flight, were loudmouths - boasting about special exploits with women as sex objects instead of people. He talked about how he had his personal lockers (in both the barracks and in the personal work area) of the aircraft decorated with pictures and clothing items from women he had spent the night and getting drunk with. Other flight crew members joined in on the conversation about the Japanese women as they came back through the cabin, saying how much they had enjoyed their company while on this voyage.

A Most Frightening Two Hours

We were four and a half to five hours into the flight when some unpleasant things began to happen. At first there was a slight jolt, then flight seemed to be back to normal except for a small change in the air flow and sound. Next, the Commander set the take seats and buckle in directives and, moments later, the lights flickered a few times inside the large cabin and then went out completely. To me, this is a very frightening part of flying as a passenger, and perhaps more so since I knew enough about the planes I worked on to make some educated guesses about the nature of the problem. What we felt in the cabin was when the engineer "feathered the prop" to reduce airflow. The crew observes the warning lights on the flight station panel and knows that an engine just went out. Did they know something before the problem was obvious to us passengers but were trying to figure out and adjust for what appears to be the issue?

We knew that a rough time was ahead with any kind of engine trouble and that we were too far out to try and turn back! The distance to Anderson AFB was still roughly 1,000 miles. This situation sure makes the navigator and engineer work for their pay! The navigator may have had double-duty as radio operator to send out communication messages about our inflight emergency and present location: messages as to the altitude and fuel load, passengers and luggage weight; lots of data that would be beneficial to the air/sea rescue unit stations on some of the small islands and to the Japanese fishing vessels in the area as well. The engineer would be reading and searching the emergency flight manual on how similar problems with the aircraft were handled in the past, while the pilots were wrestling with the controls to keep the aircraft flying as it was already losing some altitude. This KC-97 was an incredibly good and safe aircraft. General Curtis Lemay had used it as his command ship a few years earlier.

The loadmaster was doing his best to keep calm while passing out parachutes and life preservers to each of us. He hurried, hoping for time to do a demonstration of the proper way to don a parachute. It is critical to tighten the straps around the thighs as tight as person can tolerate the pressure because, when the chute opens to break the weight of their fall there is a hard jerk so, loose straps will leave a burn or do much worse should they come up into the groin. There were some (nuts) who laughed at the idea of jumping out in the ocean only to be eaten by sharks. The Airman seated near me agreed as I did that all cautionary steps should be followed to save lives.

Every man was frightened and expressing their thoughts and fears in different ways, and many comments were made about why they were even on this aircraft or in the Air Force to begin with. My prayerful thoughts came from guilt; for I, as well as many others on this flight, had sinned against God and broken the commandments. I was not trying to make light of what I did, but this event had me questioning whether this was possibly some punishment for me? "Are we to die out here in no-man's water?" I would be surprised if most of the men did not search their inner soul about faith and belief in or about God in these uncertain moments.

Iwo-Jima for a Short Stay

A sigh of relief came when one of the flight crew stated that they had contacted the rescue unit on Iwo-Jima, and we were changing course in that direction. After about 15 minutes, a cheer went up in the front cabin from the flight station - Ground Approach Assistance had our flight on radar! Then, a few minutes later we could feel a controlled descent as if we were going to land and saw vegetation below out the window shortly thereafter. We came in hard and fast on the runway at Iwo-Jima Island. Wow, what a frightening experience! We all began to rid ourselves of the life vests and started storing the parachutes in their proper places. As the aircraft taxied and was guided by ground crewman to an unloading and maintaintence spot on the tarmac, the small bushes and grass sure looked pleasant, and I can most assuredly speak a word for the group of passengers; "We are relieved, relieved and grateful to be on the ground!" When the portable ramp was being guided up to the door, the crowd of airmen rushing to the aircraft hatch as it opened reminded me of a football team running onto the field at the local stadium after a victory.

Someone in the group of airmen called out, *"Is there anyone from Alabama?"* then others started calling out their states by name like: Georgia, New York, and Chicago, Illinois. I received an unusual surprise as I heard my home state called and Thurman (Red) Neil and I made our way toward each other for a most happy reunion. God works in such mysterious ways bringing two country boys from the same small town in Tennessee together in this strange place! Only the great God of Heaven could arrange something as good and great as the joy of meeting someone out here 6,000 miles from where we were born and raised. Another thing that made me sure of His presence was discovering that the base chapel was the most important place on this island.

First order of business for the aircrew members, and two passengers that knew and understood the reciprocating engines on this KC-97, was to begin to work on the electrical and lighting problems. They were disconnecting and reconnecting wires in the hope of resuming the flight to Guam in the evening should they receive the written waiver from the Commander on Guam that was required to attempt a take-off and flight on only three engines. Most of us knew that wasn't likely to happen, but an aircraft was dispatched with some parts from Guam and a couple of maintaintence supervisors would be flown out the next day at some point.

The Commanders from the aircraft and the base at Iwo-Jima, along with the other officers and non-coms, were to make plans for the rest of us to spend a couple days on the island. At daybreak, the other aircraft arrived with parts and the news of the waiver stating that the KC-97 could depart with minimum crew, less fuel, and no passengers. Many were angry and complained about this situation. Yes, we were stranded for a few days, but most were relieved that the Air Force wasn't going to try to fly us a thousand miles on three engines. The Commander and Chaplain had plans to keep us busy and occupied. In the meantime, Airman Neil and I enjoyed our time together exchanging news from Tennessee and sharing our many other stories as to how we arrived out on these islands in the first place.

Red and I sat together as they bussed us to the Chapel and showed us some actual war films of the Marines assaulting the beaches as the Japanese rained bullets, shells, and other explosives on them. It was, at first, a slaughter until our side gained a foothold thanks to the tanks and flame throwers which could shoot and spray flames into the caves and bunkers. They also showed the historical film of the flag-raising on Mt. Suribachi and the movie "The Sands of Iwo-Jima" with John Wayne, John Agar, and Adele Moro; and some of the series "Victory at Sea" with Walter Cronkite narrating. All of these explained why it was necessary to take Iwo-Jima instead of by-passing it. There were three airfields on the Japanese-held island that would fire upon the US aircraft as they flew over or near the island. It was also determined to be a strategic location from which the B-29 bombers could strike the main island of Japan (850 miles rather than 1,600 hundred from Guam, Midway or the Wake Islands of WWII). America paid a high price to acquire Iwo-Jima with 6,821 Marines killed and 18,375 wounded. The loss of the island was more difficult to explain with 23,000 Japanese killed.

All this kept us occupied until it was time to eat and find a bunk in the barracks for a couple nights. Thurman and I continued to hang out together, talking about home and my brothers whom he had been closer friends with because they were in his class. The night was spooky with all those island birds and small creatures making unusual sounds we had never heard before. Morning came and Red Neil had to go perform his radio dispatch duties, watching for other crippled aircraft needing guidance to safety.

I developed flu-like symptoms and spent the day in the dispensary (or sick bay) as some refer to the military health facilities. It seems my body was not adaptable to the extreme change in temperatures I had experienced between leaving Japan in the low 40's and dropping into the 86-degree heat here, and this had brought on fever, chills, and a headache. My buddy Red came over to sick bay laughing at this Tennessee boy wilting under some heat. He wanted me to walk down to his favorite place and go fishing with him. This turned out to be on the end of an exceedingly long pier where he spent time catching many strange-appearing fish and other ocean creatures.

Before the sickness affected me, I took the bus from the base chapel that travels around from one end of the island to the other. The fifty-some remaining passengers and I went on a journey to the WWII War Memorial for a flag-raising ceremony. I remember being awfully glad and happy to be in attendance that day. On the return trip I must have been overcome with excitement and told a joke that I thought would cheer up some of the airmen. While it was humorous and brought some laughter, there was one black who was offended and wanted to do more than get in my face. The man was not from my shop or section; we had never met or seen each other before this trip.

The following day a C-54 came in to pick us up. We were all incredibly happy to see that plane come in and looked forward to boarding an aircraft with such comfortable seats – not to mention the four good, working engines. For the permanent-duty airmen stationed on the hot, muggy island, our visit had a been a delightful break from the monotonous daily routine of their duties so they were sad to see us go. Airman Neil and I had discussed the vast differences in our Air Force 'jobs'. Whereas I was a jet engine mechanic and my duties had basically involved removing and replacing parts, his specialist duties as a radio operator were more difficult in comparison. They could have him sitting at a desk, or seated on ammo boxes out in the field, with a headset listening to transmissions and codes on different communication channels. Red had volunteered for the one-year assignment on this island to get away from the cold weather up at his previous station near the Canadian border. I determined from our conversation that he was much more suited for his position (and the Air Force) than I could possibly be. Honestly, I was homesick and being with him did not help that situation one bit. Some airmen with other specialist jobs called us jet mechanics "blowtorch" and "stovepipe mechs" and a few other derogatory names that I do not care to repeat.

When the aircraft was said to be approaching the runway, we began collecting our gear. Those who had purchased presents in Japan were carefully carrying them out of the barracks where we had spent a couple of anxious nights. Red and I were sticking close to each other wondering who would reach Tennessee and home first; I was not due for release until next May.

Anderson AFB and the Final Month on Guam - Dec. 1957

After only 2 weeks in Guam, I had been very pleasantly surprised to receive the very nice R&R to Japan; it was something I would remember the rest of my life.

Our bunks and the barracks were mostly as we had left them. Some new troops are moving in within the next 20 days with another Wing and its aircraft to replace the Ninth and our B-47's.

Returning into the minor repair shop after eight days away, the workload had increased. Master Sergeant Harmon called me into his office to inform me about the change in store for me - I was now acting supervisor and had three Senior Airmen on my team. I suppose the total look of surprise on my face is what made him chuckle. The guys are newly reenlisted, and they want them to get oriented by someone that has been with us for a while and someone who is used to our way of accomplishing the repairs needed in the engine shop. Each shop has its own personality, special tools and equipment, and particular locations, etc. It was my hope that Harmon had something more in mind for the shop, as well as for this Airman First Class, than he laid out on the surface with this assignment. The lights that came on in my mind were that this was a test for promotion (a carrot stick hanging out) to encourage me to reenlist. These sergeants are educated in psychology, and they use it very effectively on us lower ranks.

Before beginning an assignment on the night shift, Tech Sgt. Thompson gave us a day off. The new job was simple except for picking up, straightening, and sweeping the floor at the end of our work shift; none of us cared much for that part. One of the men assigned to me became very lax in his approach to the work and the others soon followed the same route. After a few days on the job working together, the men started having an attitude problem following my instructions and with my leadership. These men were older and figured that I might be easy and allow them to do as they pleased; therefor they began to test my limits. Their animosity towards me grew by the day.

Apparently, most who reenlisted at that time either couldn't find work or wouldn't accept the pay from the jobs they were offered. I suppose these guys had some resentment toward the Air Force seeing as how, having more than enough time in grade for E-5, they had opted to re-enlist for bonus money instead of being promoted. They took their disappointment out on me because it was convenient, I guess. I tried to give them every break possible and that isn't the correct approach, but I was new at this and had soured on the Air Force because of such low morals. One of them became intoxicated and cursed me out; that was something that hadn't happened since my fight with the soldier in the National Guard at Ft. Jackson, South Carolina three years earlier. My thinking was that I had given all three of these men every break possible within my authority and, "This is the thanks I received for trying to be a nice guy?"

Storm Warnings

Knowing that there weren't many days left and I hadn't explored the island stores, restaurants, or mall out here on this island, I decided it was time to check out Guam. Later I learned there was a flea market and wondered what they could sell out here in this isolated place. Well, along with local fruits and vegetables, the neighboring islands participate by bringing their crafts. I did enjoy purchasing some items to take back to the States. (The city of Agana is as large as Lebanon, Tn.) After returning from shopping and exploring the fun things on the island, it was back to work and the news that a serious typhoon is on course to hit the Island of Guam!

These storms come periodically and are very destructive; they destroy airplanes, buildings, and trees just like tornados do in the States. Immediately, we started tying and chaining down everything that could move. The Air Force, Navy, Marine and any commercial aircraft have to be flown off this island to safety. The B-47's can go to Wake Island or Japan, and the fighters and interceptors can fly to Iwo Jima or wherever the emergency plan for this base calls for them to go. Each aircraft that is repairable gets the full house treatment. In other words, it's "all hands on deck" to fix it ASAP and, if it can't fly, it will be hangered - no matter how crowded we must make room.

The Typhoon of 1957 Sure Gave the Islands a Jolt

We had not been back on the island very long before the typhoon struck. When it hit the island, we were frightened and for good reason; this was a dangerous storm of wind and rain with rocks and gravel flying like bullets through windows and doors. All the aircraft had been flown elsewhere due to the advance warning. These barracks are built to withstand these high winds of a hundred plus miles per hour. The permanent people stationed at Anderson AFB knew how to manage the situation by moving us from one wing to another as the direction of the wind changed. The barracks were L and U-shaped which allowed the windward side to be taking the direct hit. Although the movement from wing to wing directed by the experienced barracks chiefs kept us safer, I came very close to being hit with a window shutter as it flew past my head at about 100 mph right after the first sounds of gravel hitting the sides of the barracks were heard.

The storm, which blew down the gymnasium and uprooted and took down trees, seemed to go on for a week but the worst and most damaging winds only lasted about 12 hours. We were very hungry and needed to eat before the high winds stopped blowing, so the mess section passed out sea rations and cooked some powdered eggs for two or three meals. (Perhaps the typhoon had caused all the fresh eggs to break or be contaminated, however it may also have been the Air Force's way of getting rid of those nasty powdered ones!! HA! Boy what a mess trying to clean up after such a storm.

The Sputnik Cold War

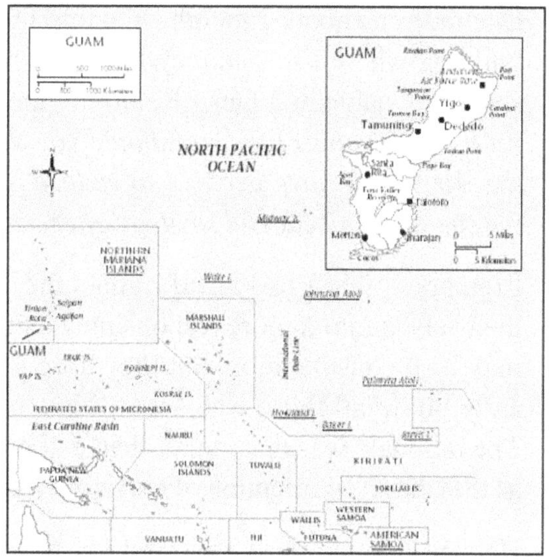

Before the storm we had enjoyed the swimming pool. That water was the most enjoyable place to be because it was the perfect temperature. After the typhoon, the pool was muddy and its surroundings devastated by beach rocks and stones. The cold war was on my mind as I stood by the pool and watched the Soviet satellite, Sputnik, pass in the sky overhead. (It appeared most visible in the days around 2 o'clock, so we would try and get a view whenever we were able.) I think there was fear all over the free world believing the Soviets could now reach any city in the United States or Europe. The Island of Guam would now become the "tip of the spear" for America's defense as aircraft could get to enemy territory quickly from there if needed. The USAF also had B-47's in the nation of Turkey and in the State of Alaska, but God is our greatest protection - not these bombers.

I once had pictures of the gymnasium that was blown down during that typhoon. The picture to the left was taken of me in front of the barracks before the repair crew came to fix the missing shutters and other damage.

While we were on Guam, the USO came over and did a few shows. Barbara White and Anita Bryant were with the group; they were so young and beautiful, and we loved their part in the entertainment. I am not sure about the time of the USO visit (more than likely in the month of November) but, other than the show at Anderson, it was not for the 9th Bomb Wing in particular. Some other shows down near Agana were probably for the other three branches of military service men and women on the island. The Air Force Base is on the higher, north end of the island. The city is more central, down by the ocean.

Guam: The Final Chapter and Trip to the States

One morning about 2 months into this TDY, I was pulled off the engine work and directed to the retention building where I'm told they have news for me. Walking into the main office, I saluted the officer and stated my name and rank. Returning the salute, the captain opened up a folder and said, *"Airman Bowe, we have a proposal for you should you decide not to stay in the Air Force. The Secretary of Defense and the Air Force have decided to reduce down some of the armed forces due the President's decision to pursue a more peaceful strategy after meeting with the Soviet leaders."*

I remarked that I had heard rumors the Air Force was getting rid of those who wouldn't try to advance themselves into a more responsible position or that continued to be a problem. The captain told me that may be a consideration, but they have a memo that goes beyond that: They were instructed to offer an early out to all those who are within a 6-month window of the end of their original enlistment contract. The rationale was that, rather than extend or re-enlist for another six years, some would opt to leave early as they have no intention of staying with the Air Force.

He said that they had reviewed my files and noted a good record with recommendations for promotion and were compelled to make an inquiry into and about my thoughts concerning this proposal. Since this was new to everyone, he went into more detail about how I could accept this offer with no reduction to my benefits or any other penalty. Continuing on, he said, *"For the record, I would like say that we want to retain you, and all the good men like you, in the Air Force."* He informed me that he was required to submit a report of this offer, as it had been proposed to me, and my decision had to be given within 15 days.

When asked if I had any questions, my reply was, *"Yes, do you know where my next duty station would be should I decide to stay with the Air Force?"* The answer to that, I was told, would depend heavily upon my decision to stay and involve an extensive inquiry. Since this was a new offer for us 'short-timers' as they termed us, they wanted to be sure about our thoughts of the future and told me they would be sending someone to talk with and help me through this particularly important decision.

A couple of days later, Dennis from the retention office picked me up in an Air Force vehicle. Typically, we would go get ice cream or something from the snack shop, then ride around for 20 minutes or so while he would point out the advantages of remaining in the Air Force, such as: the travel, a steady paycheck each month, the medical and dental care, promotions for meeting standard performance in my field, off-base housing for the family should that be in the plans, sports activities (tennis, hand-ball courts, swimming pools, and "hot houses" as we called the saunas). *"Besides these,"* he said, *"there are also other commands besides SAC and other careers besides jet engines and aircraft jobs."*

Decisions, decisions... Should I stay and receive that sergeant stripe promotion or take an early out and leave active-duty Air Force status? My original contract was for 6 years – 4 active and 2 inactive because of the National Guard time already served. (A total armed service commitment of eight years as a volunteer.)

I didn't know at the time that I was tired of the islands; homesick for the hills of Tennessee and being able to communicate with my parents, brothers and sisters, and Uncle Bernice and Aunt Mamie. Dennis was trying to be my friend and help me figure out where my mind and thoughts were, but the problem was I did not know myself.

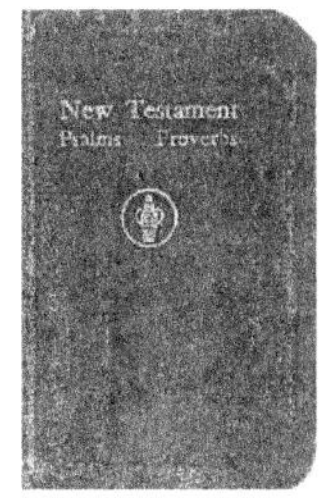

Without a doubt, I missed church: worship and bible study, church friends, and being around and able to look, talk, and laugh with the girls and ladies. I missed Miss Idaho; she had written me some good letters and sent me a small, nice carry-size Bible. Speaking of Idaho, the stories in her letters said there was snow and ice on the streets and temperatures down in the twenties, with no warming trend in the forecast.

Some of the other staff sergeants in our maintenance group had been talking with me about staying and mentioned that they enjoyed my company; both working with me as an instructor and on liberty with me. We had some unit gatherings and games together and most all knew me well - my likes and dislikes on and off the job - and had learned that I was a church person who held certain beliefs and values about the Bible and Christianity. Most, except maybe the new airmen, knew and respected my beliefs. They also knew who had given me a hard time and who had tested me almost to the point of physical action. Another thing they had learned about me was that I was not a stoolie as I had the opportunity to inform on some minor violations and did not do so.

Speaking of games played together, many of the Jet Engine Shop guys would play kickball or dodgeball and of course we would have snowball fights back on Mountain Home AFB. The last weekend I spent with them on Guam, we played a kickball game that stands out in my mind because of comments one of the airmen made when I fouled up the play on our side. This airman was a friend and made the remark, *"We will have to overlook Bowe because he is a civilian."* M.Sgt. Harmon spoke up and said, *"Civilian my butt! I will have him picking up the trash after we finish this game."* We all laughed at the irony of the statement. We knew it was basically an empty threat to point out that the Air Force still had power over me right now. I suppose it may also have been a resentful one, too, insinuating that I had rejected his unit, his shop and command.

Monday morning came and I was prepared to go to the shop and continue working, but after I had turned down the promotion and gave notice that I was going to accept the early out proposal, I now belonged to the retention or separation authorities. They drove me to the Air Force barracks, helped me pack my travel bags, then drove down to the airfield used by all commercial and military transports. We scanned the flight-scheduling board for the next available connections going Stateside. The next flight was a Navy Constellation going to Honolulu, Hawaii, with a fuel stop scheduled at Buchholz Airfield, Kwajalein Atoll - a part of the Marshall Islands chain.

That first scheduled flight was either canceled or delayed until the next day; perhaps waiting on a plane from Japan. That next flight was also delayed and I was taken, along with three other service members, to the transit barrack on the Navy side. Two of the men were Petty Officers so they were taken to other quarters. The third airman and I had the barracks to ourselves that night and were then awakened so early for the flight the second morning we had to go without breakfast.

Encouraged to keep a record of this separation travel for per-diem purposes, on the 23rd of December 1957, I wrote: We are 2,446 miles from Hawaii while in Guam, 800 miles from Wake Island and almost the same distance to Buchholz Field, Kwajalein in the Marshall Islands. The international dateline is approximately 900 miles from Kwajalein. We had time to look at it and try to estimate when we would reach the States. Some of the other service members had wives and were anxious about getting home and calling loved ones who might be concerned. (These MAT's flights do not have a set schedule.) Delay after delay with the Navy on Guam had us frazzled but we finally did have something to eat.

We boarded the Navy Super Connie and lifted off at 9 or10 PM headed for Kwajalein Island some 800 miles away. Some had expected we would be going to Wake but there must have been a VIP awaiting transportation to the States here or something. The Air Force designation for the Super Continental was either C-69 or C-121 after updates and modifications. They would cruise at 300 miles per hour. We finally lifted off and were due to arrive on the Island of Kwajalein in roughly 3.5 hours, around midnight local time on the 23rd.

Some of those pilots were so skilled that we could not feel the exact moment of touch down - no squeak, or squawk from those tires. We were invited to deplane with our carry-on bags while they refueled, and the mechanics and ground crews made important engine and fluid checks. The captain and crew chief were concerned about something minor (which is normal in the reciprocating-engine world). We boarded the C121 along with a half dozen additional passengers all hoping to go Stateside. When everyone was on board and the headcount completed, the door closed, and seatbelt lights came on. Again, the stewardess demonstrated the use of the 'May West' survival vest (a special inflatable life preserver on the back of our seats) while she also made mention of and pointed to the emergency exit doors. The aircraft taxied and went into a roll for takeoff before I even had time to finish my prayer of thanks for guidance, safety, and a trouble-free flight. Amazingly comfortable flying in the Connie R4V, I fell fast asleep.

As it was easily over two hours across the International Date Line, I woke up thinking we were in Honolulu and circling for landing instructions until the sailor beside me said we were having trouble and turning back to Buchholz Field either for repairs or to change to another flight. So, what had happened? The plane was within 300 miles of the point of no return because there is an excessively big stretch of ocean on a 1700-mile flight from Kwajalein to Hawaii with nothing in between. We were stuck there in our seats to wonder what was wrong. Are there fuel or other fluid leaks? Does the captain or a crew member have a serious health problem and in need medical help? A serious fuel leak is the most likely problem, but electrical problems are common in the Super Continentals as well. All engines seemed

to be operating normally, smooth and in sync. (We would have felt a bump or quick drop and an airflow sound change if there were engine problems.)

Due to the return to Buchholz Field, this flight crossed the International Date Line and went from the calendar date of the 24th back into the 23rd of December 1957. Interestingly, if we stay past midnight, we will be into the 25th - my birthday. There must have been a fluid, oil, or fuel problem because the captain and navigator calculated that the plane couldn't make the other 1175 miles without some strong tail winds, which they didn't have. Safety for the passengers was the determining factor in returning. Even if the leak had not worsened, a head wind was forecast most of the way to Hawaii and even at reduced power this would leave no safety margin to travel the remainder of the way. We landed safely back in Kwajalein where we had departed from 8 hours earlier. The crew received a hand clap and cheers for good performance and making the correct decision to return.

They changed to a well-rested crew and gave the passengers a choice: waiting an hour or longer while the fluid lines were changed to continue on this plane; or waiting possibly six or more hours until another MAT's (Military Aerial Transport) flight arrived from Guam. About one-third of the passengers deplaned for other arrangements so the ground crew took the extra cargo and some of our luggage off. Apparently, there is still something wrong this C-121 that needs to be repaired because they plan to stop in Hickman, Hawaii, before continuing the remainder of the 1700 miles. If they can cruise at 325mph, we should make Hickman in 3.23 hrs.

Traveling light and fast, we made the trip to Hawaii (where it was still the 23rd of December) without incident, but it was the 25th back in Guam. No time to sightsee in Hawaii this time because the next MAT's C-124 is taking off for the States within the next 2 hours. I am very tired and anxious to get back, and into a bus or auto instead of these airplanes.

The Loadmaster and a huge amount of cargo was all that occupied the plane when I boarded the C-124 headed for California, USA. These large planes fly considerably slower than the C-121's. They are manufactured for a different purpose with enormous space for heavy loads of supplies and vehicles, like trucks, tanks, and bulky freight pallets.

Once we boarded and strapped in, with all the noise and vibration as the aircraft headed down the runway, I thought, "Boy do we need a touch of the master's hand on this flight!" Yes, I still prayed as always on takeoff - that is the most dangerous time because, should something go seriously wrong, there is no turning back. Just hope they get the aircraft high enough to make corrections or try to figure out the problem and look for a place suitable to land.

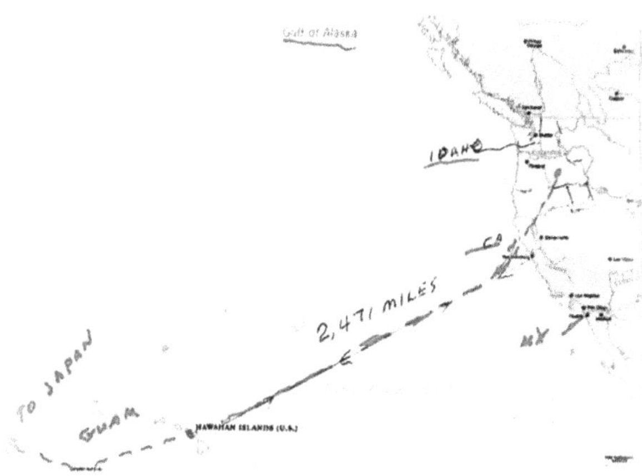

I am slowly getting closer to home base; these 2,470 miles will take another 12 or 13 hours before we step onto U.S. soil in California. As is the case with any aircraft, they must fly in all kinds of weather. The weather report and conditions can and often do change within a few hours. This MAT's flight hit a storm they could not fly around, so the Commander will have to decide to turn back or choose a way through the storm. He does this by using radar and trying to pick the way through the freest spaces in the clouds.

What a rough, bouncy ride this turned into! No sleep or rest on this flight during 4 hours of what seems like riding a bucking mule out on the farm. Most mules we worked and broke or trained to perform the farming did not try to throw the riders as horses do; but occasionally, Dad would acquire a young mule that tried to buck me off. If we weren't strapped into this side-facing foldup seat by a shoulder harness, the loadmaster and I would be bouncing off the ceiling - it's that bad! Then of course sickness and vomiting follow all that violent moving up and down from side to side and there is nothing in the galley that would help even if I could unstrap and go there for some. I do not get car sick like some do, but this?

> Changing subjects: Flying so many hours without being a crew member is common for Air Force personnel. People in the SAC command will fly as passengers with the Navy transports across the oceans while those in other branches of the military may fly commercial and/or with the Air Force MAT's. There are some necessities built into all aircrafts, especially the cabins, depending on their purpose. The C-54, C-121, C-47, and the C-119 are primarily passenger aircraft that have the rest rooms and the galley (kitchen) in the typical rear of the cabin location. In the galley there is a small refrigerator and coffee maker which the flight attendant usually takes care of as part of their job; but on some flights that do not have scheduled attendants we serve ourselves under the watchful eye of a crewmember. These long flights have a secondary crew and extra food and drink provided. Basically, they have refrigerated sandwiches and drinks: coffee, tea, small boxes of milk or juice and soda.

The C-124 finally flew into some smooth weather and I fell fast asleep for a few hours. The body can only take so much shaking and bucking until it becomes so tired that it will sleep through most any and everything. The best news was hearing we were landing in the good old USA.

Active-Duty Separation

When making decisions about what your next career move should and could be, who do you talk to? Who do you trust? I should have been praying more but I was homesick and did not understand my emotions. One decision had already been made, giving up E-5 or a possible promotion to Sergeant for an early out. As the Apostle Paul said, "forgetting those things behind and moving forward." I guess Will Rogers said the same thing. Looking back, my decision to leave the Air Force was the right thing to do long term, but the next two years' struggles were painful because money was short and all the jobs were low-paying.

While both of my brothers were on "flying stats", which added at least $50 per month to their pay, J. C. and Bud were also locked into a frozen promotion and pay grades field because of their AFSC's (Air Force Specialist Codes). On my side, as a jet mechanic, the rank and pay grades were wide open in this "New Frontier" of the Air Force so all my promotions came on time. As soon as the minimum time in grade passed, I received a promotion on the next cycle of orders from Wing or Squadron Headquarters every 6 to 9 months. There could be several open slots on base for E-2 through E-8 and the Master Chief would place all the eligible names in his command for promotion, unless there had been misconduct by someone. My date of rank E-4 was Dec. 1957. My 48 months active duty wasn't supposed to conclude until May 17, 1958. I turned down the E-5 promotion in Nov. 1957 and was released Jan. 9, 1958, after serving proudly in the Air Force for 3 years, 7 months, and 21 days. I served with a great group of highly dedicated men.

Back in the USA on December 24, 1957

From Castle AFB in California, I caught a bus to take me back to Idaho where I would be released from active duty and enter the Air Force Reserve in which I remained until May of 1960. My route took me through Northeast California and Nevada, celebrating my 23rd birthday as we stopped in Winnemucca Bus Station, and then continuing to Mt. Home Air Force Base. Before leaving out for the TDY to Guam, I had left my car with Minnie, so she and her mother picked me up in it and we went straight to the base so I could sign my separation papers and gather my personal belongings. They took me back to their house afterwards. (During the car ride, I found out that Minnie had told her mother I was going to marry her when I came back.)

The time in the military goes so slow on a four-year enlistment. I had initially dreamed of a wonderful career in the Air Force, but most of that changed when I became aware of who I was: a Bible-believing Christian. Coming into the service was so much about growing up and learning what were the most important things in life – and it was not the military. The awakening about church and belonging to the Lord God at the testing center and being baptized at Mountain Home changed my outlook so much.

While Minnie Dew was a potential bride, many elements had to fall in line or place before we could think about or discuss marriage; primarily her graduation from high school and me finding a job in a career with a future. With thoughts of exploration and adventure still swimming in my head, being free to be all I could be as a single young man, I knew there was some growing that needed to take place. I left Idaho in January of 1958, and a family needing a ride accompanied me the several hundred miles to Denver, CO.

When this couple and their small son and I departed Mt. Home in cold but fair weather, the road was clear; but boy did that change soon after going through Twin Falls where the Mercury spun out of control and off the highway twice in the foot of snow. Although we were able to push and shove and recover onto the main road quickly, the lady became very nervous with my driving and insisted that she would feel safer if her husband took over. I let him drive several miles until we reached a lower elevation which was free of the snow cover. However, as we traveled through the mountains of Eastern Idaho and Western Wyoming, the twisting and turning and weaving side to side through the crooked mountainous roads soon made the young child very sick and he threw up in the car.

In response, the stepfather stopped the auto, took his belt off, and whipped the youngster hard. Neither the child's mother nor I had the courage to defend him. (This is something which I have lived to regret.) We traveled on through the mountains of Wyoming through Cheyenne and on into Colorado, finally reaching the struggling married couple's destination of Denver. While I was grateful for the company and assistance driving those treacherous mountain roads, I was also thankful that I didn't have to listen to the arguments and the blame game of these two failures as homemakers any longer. (They certainly needed Christ in their lives!)

Before leaving Idaho, I had started reading about careers and schools that offer a job while a veteran used his or her GI bill to help pay for the schooling. North American Aircraft Engineering School in Tulsa, Oklahoma, caught my attention. Since I had spent three years working on aircraft, perhaps there would or could be a future for me in the aviation field. So, my plan after dropping the couple and child off in Denver was to head for the Spartan Aeronautical School and the North American Aircraft Manufacturing Company there. After reaching Tulsa and the school, which had very nice campus and impressive surroundings, to my disappointment there had been a downturn in the economy and they were laying off the work force and could not offer the school students a job. So, God shut the door on that career possibility quickly. (Whether it was necessary or not, the President had cut funding after meeting with the leader of Russia in order to pursue a more peaceful atmosphere with the Soviets.)

Well, with that idea shot down, the next stop and destination was home to Tennessee. I was homesick - didn't know the symptoms but homesick. It was going to be a while before the Lord would open the other door to what I was more suited for: Operating machine tools and using my God-given mechanical ability to learn the trade that was intended for me all the while. I had tested out highest in mechanical and electrical and technical skills.

Yes, I could not wait to arrive home and rest up from the snowy and icy roads that had transformed me into a bundle of nerves. I suppose the flying from the islands to the States, in bad weather especially, changes the sleep cycles from day to night - they refer to the difference as "Jet Lag". The 1700-mile drive home had me on pins and needles so badly that, when I stopped to spend the night in a motel in the state of Missouri, I could not shut my eyes for sleep. I was a wreck!

The 1954 Mercury stick-shift was an unusually nice auto; it not only got me back home to Tennessee but also served me well for the next couple of years. After being somewhat restricted to my job and the base, now I had some freedom and a short relief from the routine of getting up, putting on a uniform and going to the mess hall for breakfast. Arriving at my parent's house, with the boys and two girls gone, it was such a different place now. Mom, dad, and Shirley were such a welcome sight! And the bed, boy what comfort; I never remembered it feeling that good when it was available each night before I left home four years earlier.

Here at Mom and Dad's place with younger sister Shirley for a few mornings, it was going to be Mom's gravy and biscuits (which I had surely missed these past 3 years and 8 months). It was January and cold, not nearly as bad as the Idaho winters, but it was also a different kind of cold. During this reprieve, some of the remarks people had made while I considered whether or not to leave the Air Force kept ringing in my head, *"Dorris you will be in the soup line if you leave us; you will never make it out there."* The world economics and job market were such that these statements were echoed from both peers and supervisors that I served with in Guam and at Mt. Home. Friends, co-workers, and NCOs wanted me to remain in the Air Force with them.

The fact was, if had I re-enlisted, I would have been shipped overseas to one of several places: Turkey, Saudi Arabia, Alaska, Japan and possibly even Korea. This was another reason why I no longer felt the military was a career for me. Although I had done well, not great but well, the recommitment I had made to the Lord Jesus when I entered the Air Force was just not consistent with that lifestyle. That idea did not fit my desires at all - a wife and family seemed more suitable to this hillbilly. I was ready for another change. Money and a job also seemed more in line with my thinking.

Potential Wife Number Two

Next on my agenda was to try and find some old schoolmates and friends. Earl and Micky, and perhaps Robert, were living in the hometown area. Cousin Jack Bowe was living and working in town at the bank. Wanting to welcome me home in a special way, he had me meet him and his lady friend with a promise to introduce me to a special person. When they came by to pick me up, I was surprised that I knew the lady that was going to be my date for the evening. We had known of each other because we had played sports at the same school but never spoke to each other except in passing. She was four or five grades back in Jack's class, so she was in grade school when I played football. A basketball star, I had watched her play from the bleachers while home on my first furlough.

Pepper Draper was very attractive with short, blonde hair and a nice trim athletic body. She became my closest friend for the next 3 months. But what about Minnie back in Mt. Home? I could not feel quite right and comfortable with this new lady friend. Many times, decisions come slow, but they must be made. What will I do with two potential candidates for wives yet no work to be able to support a wife and family? I knew I could draw unemployment after my severance pay expired but that was not the answer. That was going to get me "in the soup line" as the Air Force men had predicted.

Now the job search was on but there was nothing local to be had. I decided to try Nashville; since it seemed there were always open jobs for experienced mechanics, perhaps one of the auto garages was hiring. Besides, I couldn't draw unemployment without applying for some jobs anyway. Having worked a lot on my Ford out in Idaho, I knew enough to get me in the door, so I took a job as an auto mechanic. The pay was minimum wage plus commissions, and I soon realized what the airmen meant - it takes a lot of money to live on when the government isn't paying the bills. Now I understood why my brothers had reenlisted. After making the trip back and forth 70 miles a day, it seemed more of an idea worth considering now that I had to pay rent and keep the auto insured with good tires and out of ditches. Am I going to have the same chance to earn money to support a wife and children if I don't reenlist?

The Air Force recruiters had obtained my address and phone number as filing for the unemployment had supplied the information to them also. In the meantime, I had gone to work in Nashville for a short time at a garage with a misleading name like 'Auto Engineering' that was nothing but a sweat shop. Knowing how to take an engine apart was one thing but getting it back together and having a smooth-running car took more experience than James Mason and I had acquired working on the Ford. For a couple of weeks, I lived in the upstairs of a thin-walled house in Nashville with friends from school who had nice paying jobs. Deciding I did not fit the type of environment they had accepted, I said thanks but no thanks to the 70 mile back and forth drive to Nashville on the weekends. Had I fallen in love with Pepper and had a good paying job it might have made sense, but she did not understand my thinking about the lack of a job and a future career in the area. I decided to take what pay I had earned and look elsewhere for something more promising.

To add to the pressures, a recruiter from one of the nearby cities was hot on my trail. Hoping he had something new to add to what I had already heard repeatedly, I gave in and agreed to meet with and listen to him. My brothers had reenlisted for bonuses, but they were married and on flying stats in more critical jobs. I was surplus, but with good character and record, so I was desired. Pepper and I met with the Air Force sergeant at the Cumberland Café in our hometown. Was there anything new? A stateside base! Since her brother was an Air Force pilot, she knew a great deal about the military life. He was very professional and did a nice presentation which she soaked up very intently. Hearing about the base at Dayton, Ohio with its housing for married couples and all the wife's benefits such as: the clothing, supply, and grocery stores in the base exchange, medical facilities, and more impressed her greatly. Of course, I wasn't interested because I was sick of the Air Force and the engine repair job. Pepper even asked about travel as a spouse; she may have visited an Air Base where her brother was stationed, and his wife lived there.

We left the meeting with a signed statement from the sergeant stating that he had met with me and had explained the rights and benefits if I were accepted back into the service. As I promised to think about returning before my time ran out on my pay grade E-4, my thoughts went back to rejecting the E-5 Staff Sargent offer. My other thought was, "Do I really want to marry this woman who had not shown very much affection?" When she had asked, *"Don't I get a ring or something?"* I realized she was asking me to show I cared for her and wanted a commitment. I do not remember my reply; I believe I was surprised and just smiled and stared at her as of course she smiled back. We let it go with those smiles, kissed, and went our separate ways since she had to get to work. At the school, she was the assistant receptionist recording grades, etc. Pepper was a very smart girl. She was college material, proficient in typing and all the office skills taught in high school.

> I'm sure by now that relating this portion of my romantic life probably spurs some questions like, Is this a man or mouse? Did he deliberately leave out the sexual encounters? Was he gay or what goes? Being most respectful of the ladies I have written about, although there were strong desires to lay in the back seat and go all the way, our home and school teachings were very plain: God knows everything we do. Fornication is very plainly spoken against in the Bible, so we heeded the warnings of the consequences for unmarried sexual activities. I had participated in a one-night stand in Japan and the results were that I now felt impure and undeserving of a virgin wife. I left Miss Detroit, Minnie Dew, and Pepper Draper as I found them: pure and clean.

There was an incident with Pepper that she might not mind me relating. We were parked out in the dark by the cemetery, hugging and kissing as young couples do, when she pulled away and asked me point blank, *"Did you have sex with your girlfriend in Idaho?"* And I said, *"No! Absolutely not!"* This and the story about Minnie and I misbehaving somewhat (where she got down in the front seat floorboard and prayed aloud for both of us) are examples of the extent of my romantic encounters. Even when you have the utmost respect for the other person, things sometimes get out of hand when the desires become too great.

Well, Pepper threw more ice water on the parking in the auto situation by suggesting, *"I have a couple books you should read."* She was right of course but I wasn't ready to hear that, so she tried to cajole me by telling me that her desires were as strong as mine, but she exhibited more will to do right. I suppose I started the Mercury and took her home. Many questions were on my mind such as: Do I really want this woman to be my wife? What other books would she want me to read when the situation came up unfavorable to her liking? What are my true feelings about this inexperienced, sheltered 18 yr. old? Was I the first man she had become interested in? At that point in our relationship, I doubted my intentions toward her.

We continued to see each other and attended the Methodist Church in town where she played the piano - beautifully I might add. I took her to meet my folks and, not surprisingly, they liked her and enjoyed the company. Mom may have felt awkward but made everyone that came to our home very welcome and comfortable in her presence. I was cautious about taking visitors to the home place. While the house was clean, it was piled up and poorly decorated. (This was my problem, not the folks' - they were doing the best they knew.)

Pepper enjoyed driving my Mercury because of the stick shift. One day, she ran it into a ditch while taking some of her friends for a ride. All nervous and apologetic as she came back to the restaurant where she had left me, we inspected the auto for damage and there was none. I drove slowly and was cautious when we went driving on the graveled roads, but when getting back to the pavement I enjoyed showing out just a bit and burning some rubber for fun. She scolded me for doing such.

About the Future

Back at home I was having conversations with my parents about what to do with my life as the money was slowly evaporating. I asked Dad if he wanted to get into farming in a more serious way since he was only 44 years old (still 18 years away from drawing Social Security) but he wasn't up for that; not with me at any rate. Interested in my wellbeing, Mom had cautioned me about being a man on the run much like her brother Thurston, who had sort of become a "hobo" for part of his life. He would leave home and his wife without any information about where he might be reached and show up months later; usually with money he had earned as a farm or ranch worker out west. Our conversation triggered the idea that, with him and his wife Dorothy living in Warsaw, Indiana, it might be a worth contacting them. This seemed like a good suggestion, so a quick call was made to my aunt to let her, and the rest of the family know I was home from the Air Force and hadn't had any success finding a job here in the Upper Cumberland region of Tennessee.

While she couldn't give any assurance about the job market up there, she said they would sure enjoy seeing me as it had been such a long, long time since they had. She also said she would inform Thurston and the girls to get the guest room ready for their nephew, Dorris. This idea came from out of nowhere and turned my thinking and planning all around. Mom and the girls had been up north to see them some months earlier on a job hunt without the results they wanted, but the family visit was a very warm welcome. There had been no mail or phone calls from my job applications in Nashville. General Nunnally, a man from my town who was working with the government in some capacity at the state capitol, had met with me about working for the Air National Guard and at the Voltee Aircraft Manufacturing Factory there in Nashville, but that wasn't panning out.

Somehow, I finally broke the news to Pepper that the Air Force was in the past and that I must seek work out of state. I knew she was disappointed and was very quiet the next few times I picked her up for a ride or something to eat out at the Dipsy Doodle or George's Crossroads Restaurant. Pepper's mother was always nice but seemed suspicious of me. There was no man around the house, and I don't remember where her father was.

Jess Bowe's Ill-fated Flight – 1958

My youngest brother, Airman A/2C Jess C. Bowe was a Second Mech/Flight Engineer-in-training on the largest aircraft in the Air Force - the B-36 Peacemaker. With an amazing 10-engines, this plane was used as a deterrent force in the Cold War and had lived up to its name by helping keep the United States safely out of other conflicts. J.C. was qualified as Crew Chief and Flight Engineer (F/E) on the B-36 and other multi-engine aircraft. He had the most time in grade there in the squadron for Airman First Class on Biggs AFB in El Paso, Texas but could not receive a promotion no matter how well he performed his duties. However, the Squadron Commander promised him a promotion if he would volunteer to transfer for a year to Goose Bay, Labrador. As he departed Biggs, he was looking forward to being a mechanic on the B-36 bomber there. Before touching down at Goose Bay, the aircraft he was a passenger on circled the field and to his amazement all the aircraft there were KC-97's (refueling aircraft tankers).

He began his assignment as a Flight Engineer: One who monitors all the gauges for performance, fuels the plane, starts the engines, and does all of the mechanical checks/2nd Mech on the KC-97's.

A few weeks later his crew went out on a refueling mission and, right after unhooking from the plane they had just filled, a light came on the monitoring panel indicating that there was a fire. Knowing the history of these aerial tankers (They explode in midair!) this crew, after reviewing emergency procedures, decided staying with the crippled plane would be unsafe, and there was no place to land within a couple minutes, so they jumped out of it. After or during a fire of any magnitude, this aircraft - whether fuel bladders/tanks and reservoirs were loaded or supposedly empty - are extremely dangerous for the crew and passengers if any are on board. Passengers in this case would have been issued a parachute just like the crew. We could not find the mishap in the archives, so as a writer I had a Q&A exercise with my brother about the incident to this result:

- My brother was the Flight Engineer and was the first to notice the engine trouble light
- He stated that the first indication of trouble was the warning light for a runaway propeller!
- The engine and propellers are on the same shaft. The propeller pitch and angle blades may be changed and reversed, as desired from the aircraft's flight station.
- There are governors controlling the propeller that keeps it in sync with the engine according to the throttle setting
- In this case, the governors malfunctioned and allowed the engine to over-spin and overheat the shaft bearings.
- Soon afterwards something must give because the oil that lubricates the bearings catches on fire; the fire quickly spreads and boom - no aircraft or crew left to tell the story!

J.C. had had twelve months training as Crew Chief/Flight Engineer and many flights as the First Engineer. It was the captain's decision, after quick discussions with the Navigator, as to where they were in relationship to an emergency landing strip and there were none. They only have a window of about 20 seconds or so before bailing out time after they try to extinguish the engine fire. The prop is likely to separate from the engine and go into the aircraft and fuel cells, exploding the aircraft.

More Q&A revealed the following:

- SOP (Standard Operating Procedure) is for the Boom Operator to jettison the survival gear kits. Then they are the first person out the door and 'into the silk'.
- The Navigator, then the Crew Chief (in this case J.C.), and then the Radio Operator jump out next.
- The Pilots set the auto pilot and activate the SOS boxes before exiting the plane

Upon leaving his station with the chest pack parachute, Airmen Bowe was aware that he had to count to ten before pulling the rip cord and make sure he cleared the aircraft's rudder stabilizer. Once his chute opened, he saw the rest of the crew – some above and some below – and some tall pine trees directly beneath that made him wonder how far off the ground he would lodge in those trees.

He continued, *"I also tried to keep track and notice where the other crew members and the survival kit chutes might be landing in relation to everyone. Fortunately for all of us, the wind pushed us and the survival kit to the ridge top, into small shrubs and bushes so we were able to locate our survival gear and each other."*

Now, to survive until command misses this flight after it doesn't return on schedule. No doubt there was an SOS message transmitted out, but the abandoned aircraft flew on out to sea and exploded or just crashed into the ocean. The search and rescue efforts would mostly be directed in the area of and centered on the last transmission unless aircraft debris was discovered out in the ocean where parts of the wreckage showed up.

When asked, Jess stated that the flares the crew fired did not attract the attention of any of the numerous planes that flew over until a civilian air carrier finally spotted one about 15 or 16 days after the jump and encampment. When asked about camping and eating, Airman Bowe replied, *"There was a tent and sea rations in the survival package, along with a .22 caliber short-barrel rifle."*

More questions revealed that they had landed near a river, but also close enough to see the ocean from the ridge top. They ate sea rations daily, but two of the officers who were from Florida did a fantastic job catching, flaying and cooking some fish for them. Surprisingly, fishing gear is not part of the survival kits, but they had improvised with the equipment that was included in the package and he said the fish just jumped up to grab the items they used as bait. At one point, when a moose came down to investigate them, Jess had begged the captain to allow him to shoot it for meat and a change in the daily menu, but he wouldn't allow that. When I inquired about the climate, he said, *"Those flight suits are not very warm or comfortable; they are tough and durable though. We had great weather for Newfoundland; it was summer and warm but not hot in July 1958."* One final question, *" Seventeen days is a long time to be away from civilization, were you afraid of not being rescued?"* J.C. replied, *"No, I didn't think much about that; being an 18-year-old, the youngest of the crew in good health and with older officers who had been in tight spots before, put me in great company. I was having a good time with very few responsibilities."*

When they took off on the refueling mission, they had the best weather report possible in Goose Bay Labrador. By the time the crew was finally rescued, they were all unshaven and dirty from the falls they had taken while clambering down the mountainside to reach the river - for fresh water and in hopes of encountering local hunters or fishermen.

There were stories that circulated to serve as a warning there at Mt. Home AFB, Idaho. One was about an airman cleaning the inside of the KC-97 tanks who became overcome with vapors and died before the medics could get the poison out of his lungs. Another was the story of two other airmen who both died drinking the water-alcohol solution that the B-47 bombers use for takeoff assistance.

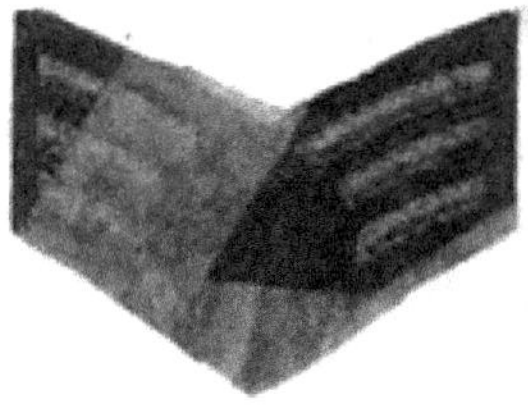

My youngest brother, J. C. Bowe, was promised a stripe if he would go to Goose Bay Labrador for a year. He thought he would continue as a crew chief of a B-36 bomber, but he was instead assigned to an aerial refueling aircraft.

The KC-97 he was on had an engine fire shortly after refueling another plane and all the crew had to bail out and survive in the wilderness for 17 days before being rescued.

J C was concerned about getting his parachute caught high in a pine tree, but the wind caught the chute and pushed him to the ridge top. He had a few bruises— the chutes usually

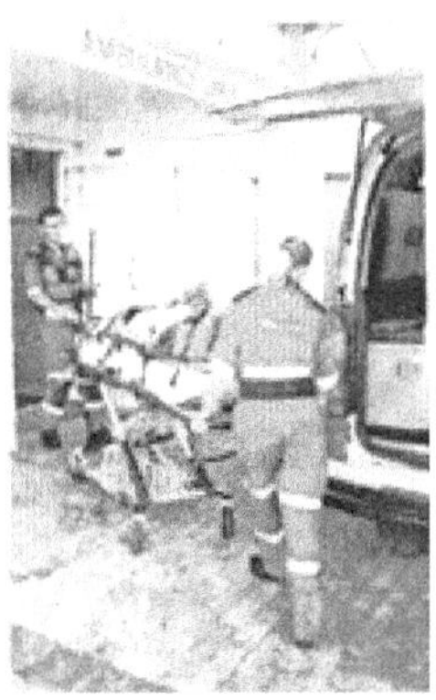

The crew of course required examination after

JC and the B-36 Bomber - Biggs AFB

In the forward of the Encylopedia of American Aircraft, Leiutenant-General Ira Eaker, USAF-Retired, wrote these facts about the huge 10-engine bomber with its six propellers and four jets: While no American aircraft had received more criticism than the B-36 Bomber, it was dubbed a "sitting duck" and a "political tool", but General Eaker called it the "Peace Keeper". In Washinton, D. C., the Navy wanted funds for a Super Carrier. The Air Force, a "new kid on the block" so to speak, contended that the Navy had defeated all the major navies of the world. While this was true, it was known that the enemy is constantly planning to be the strongest and most ready nation on the planet.

The Strategic Bomber Project or '*The Magnesium Cloud*' as the gargantuan B-36 was sometimes dubbed, demanded a whole new concept of logistics and maintenance support and was responsible for the development of new technologies in navigation and radar bombing in the rarefied air more than twelve miles up to outmanuever most contemporary fighters. Shear size was it's most distinctive feature, which led one pilot to remark, *"It is like sitting in a bay window and flying an apartment house."*

My brother, A/2C Jess Bowe, could remark on the B-36 aircraft from some personal stories and experience with it while he was stationed at the home base of one of the squadrons in El Paso, Texas. He spoke of the fuel and oil leaks that kept it grounded for days and weeks as the maintainence crews doubled-up to get the plane airworthy. The one story he related to us seemed like a nightmare as he retold observing from the ground while fire trucks and crash crew raced to their positions to extinguish the would-be fire upon the B-36's crash landing. Someone noted that the crew was bailing out as those small white billows began to appear with their airmen dangling from the open chutes. In the meantime, the aircraft landed in almost two separate pieces with the tail-section skidding along the runway while the landing gear held up the front and central portion of that huge aircraft. There was no report of what kind of training exercise or experiment had been going on several-thousand feet in the air to cause the fueselage to break into two separate bodies. (Special maneuvering or strong turbulence often occurs on these missions.)

That 'bomber' was the most controversial of all American aircraft that flew - it never fired a shot or dropped a bomb in combat! No American aircraft has been more widely critized than the (B-36) Global Bomber, and none excepting wartime achievment has justified itself more fully. From it's inception in WWII until the end of its operational career and beyond, it was the center of much bitter controversy. Its critics called it a "political tool" while others deemed it a "sitting duck" because it moved so slowly. Others avowed that the B-36 did basically all that a piston-driven aircraft could do; and did it longer than anyone believed possible. The B-36 was the first major weapons system of our time to accomplish its purpose: i.e. to deter attacks against the American Homeland from an enemy with the capability to reach the USA with its weaponry if not detected and destroyed.

Getting Serious in Indiana

Time passed by rapidly over the next few months. With the mating game on my mind, I worked some uninteresting jobs while girls, fun days, and good times occupied my life. I was living with Uncle Thurston and Aunt Dorothy in Warsaw when Linda and I met in July of 1958. The sad part is how much fun I was having being irresponsible. There was still a girlfriend in Idaho, and one I had just left in Tennessee, but I was enjoying the best worldly time of my life doing what the Bible warns about - the lust of the flesh and the pride of life.

Linda was more aggressive than any of the girls I had encountered before. We kept getting closer and closer until finally my faith faltered to the point where I decided that marriage must take place. Linda planned the wedding (Oct. 24, 1958) and I made plans to go to trade school on the GI education bill. This was the beginning of my machinist career. The idea had come from a buddy in the Air Force - Sammy Blain from Detroit, MI. He had heard of a nice salary which could be earned in the auto manufacturing industry's machine tool and die skilled trade. He did not know what a toolmaker was, or what they really did, he just knew or heard that they made good money.

We traveled to Chicago and found an apartment on the north side. Having never been away from home, and now living with a man she knew very little about in a place she knew nothing about, this had to be such a trying time for Linda. All I knew to do was go to work, earn some money, and focus on learning the skills at the machinist school I was enrolled in. I rode the elevated train to my classes at the Greer Shop in the mornings and worked nights at Union Station handling mail bags. This was a very labor-intensive job, lifting and tossing heavy double sacks from one cart to another.

> That train station in Chicago was a big deal then (in the 1950's) before the airline industry began transporting any and everything including mail. Trains and railcars were, always will be, and continue to be important and universally used for extra-large and heavy items, and of course passenger service. (Air travel, which was much faster than by train, was just coming into being when I was working the mail and learning a trade.)

Linda knew about grocery shopping and cooking, but with no one around to talk with while I was away at work and school, the isolation and homesickness were a nightmare for her. She was used to having her mother to talk to. The truth was that we hardly knew each other. Having met only 90 days before saying our marriage vows, here we were stuck together in this big city with no help or family within almost a hundred miles. I am sure she felt the same as I did about the marriage situation – thinking "What have I done? I was not ready for this!!" but this is my life and somehow, we must make being together as almost strangers work.

Our apartment was near the "L" line (the elevated train cars that run on rails all over the large cities). We had to park our 1954 Mercury under the tracks which ruined the paint on the car because the train dropped hot sparks and residue onto it from the electric shorting on the steel rails. I pulled the mail for a couple of months until a machine shop job became available. Thinking this would help advance my skills and career, that first machining job proved to be something I was not quite ready for. There was obviously much more training that I needed; but it was good for starters to show me where I stood and what level would have to be reached.

Next, I went to work for the Howdy-Doody Popsicle Bag Company. That was such a fun, but nevertheless dead-end, job. The work was easy after I mastered the skill and technique. There were several machines in a production line that I was responsible for keeping operational and in sync. These machines folded and cut off the wrapper bags. Periodically, they would get out of time and either cut the bags too long or too short and jam when they tried to cut more than one bag at a time. There were a couple other men that did the same job and I had lots of fun with the flirty women who kept me laughing by picking at and teasing me.

It was after about the seventh month in that Chicago apartment that Linda decided she could not take any more and went home back to Leesburg and the easy, happier life. She left me on good terms. Now in the advanced classes of the machine shop, I stayed and worked another six weeks with the bag company before dropping out of school around June of 1959 to follow her. I believe my father-in-law, Howard Bibler, thought well of me and he saw that I was not afraid of work. Wanting to support his daughter and save our marriage by keeping me closer to home, he found a job for me back in Warsaw at the Acme Machine Tool and Die shop. This was the place where Howard took his broken or damaged parts from his job to be repaired. He worked for Fiberfill - a glass pellet manufacturing company.

Linda and I did remain married and, over the next 5 years, we had two sons: Norman and Keith. I continued to be in the Air Force reserves until May 1960. We moved to California but my wife missed home, so we returned to Indiana until circumstances and personalities led to our eventual divorce in 1989. As a civilian, I took a job with the Navy as a machinist and I would eventually join and serve and have many more adventures in the United States Navy Reserves, but that will be the subject of another book.

Aircraft Associated with the Bowe Brothers :

Upon our enlistment in the Air Force, we brothers found it hard to believe we would end up flying so many hours and could talk for hours on end about the dangerous missions we flew—even Ray as a passenger.

All of the Air Force aircraft pictured on this page were the types they maintained, were crew members of, or that transported them between duty stations.

A C-119 Flying Boxcar used in the Dragnet Program

Buddy H. Bowe was a pole handler and injured when his C-119 crashed

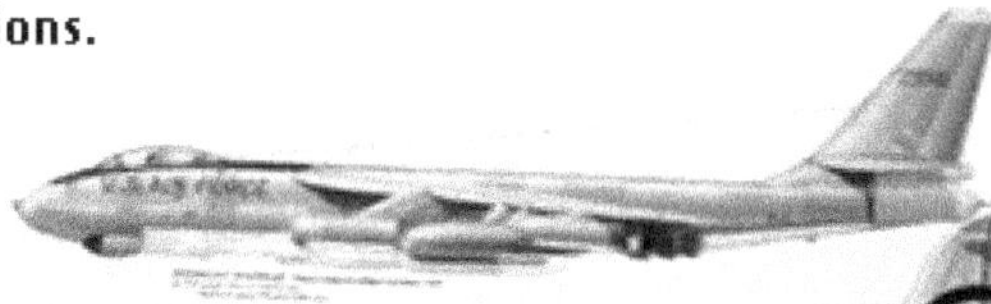

B-47

The B-36 was the largest aircraft used by the Air Force

General Curtis LeMay used this C-97 type plane for his command center aircraft

J.C. was the Flight Engineer on the CK-97 in Labrador when the crew bailed out.

The C-124 was a giant transporter for delivering large equipment and heavy cargo. Buddy became Loadmaster on such.

The C-54 was the workhorse of the Berlin Blockade and was the very first plane that I, Ray Bowe, flew on after Basic.

The Navy Constellation was the aircraft for my flight from Guam to the other islands on the way to Hawaii in December of 1957.

THE C-47 was a work horse during the invasion of Europe and Market Garten Operation when it dropped the troops into France and other places during and after WWII. We in the 9th Field Maintenance were transported to Tucson, AZ for the bombing competition in this aircraft. The Army Air Corps modified it into the DC-3 which was a very useful aircraft.

Tennessee, Oh Tennessee!

The men and women of Tennessee never shirked their responsibilities to the United States. In the tradition of John Sevier and Isaac Shelby, of Andrew Jackson and Sam Houston, of Jim Bowie and David Crockett, of Alvin C. York and Albert Gleaves, of James Davis and Frank Kelso, thousands and thousands of nameless Tennessee heroes will always be there when the call to duty is issued. Those I knew and/or served with are described in the paragraphs to follow and exemplify the same heart and spirit, therefore I wish to pay them tribute in this book. (Names are changed where necessary)

Corporal James Song: It is a pleasure to write a story about a fellow veteran regardless of the branch they served in or with. A friend and teammate, it was an honor to know and practice and play football with him at Bixby Seminary High. Many members on the 1950-55 squads went into the military and served out Dr. Harry Upperman's and Bixby Seminary's Motto: "Loyalty, Truth, Honor, and Service" and James became the strongest and purest example of that in my opinion.

Corporal James Song was in the toughest branch of all the services: The United States Marine Corps. They were trained to be the best fighting men America could produce. The USMC is one of the proudest and most impressive branches of the United States Armed Forces. In and out of combat, Marines have stood proud and brought honor to this country. (The Few, The Proud, Marines)

Of all the people in our class of 1955, James Song would have been the last one I would have thought of to go into the Marines on account of he was so quiet that hardly anyone knew he was around. What I knew of him came mostly from our participation in football. Our team had more backs than linemen and the two groups ran and practiced separately and took instruction from different coaches. James had speed but exhibited an unusual manner and unique style of running: His upper body and head were not down and forward as most half-backs carried themselves, but rather James ran with knees high so as to ward off tacklers. We did a lot of snickering about his head back, upright style, but what does a lineman know about the correct posture for running backs? Personally, I did not like what was referred to as 'Head on Tackle' (one-on-one attempts to either tackle high or knock the football loose) but James handled that with fun and ease. We would charge each other from five yards away.

If I had any classes with him, it would have been shop math. He took the building trades classes and was busy with such things as the construction of dummy structures or the frames and foundations of buildings. Our personalities and friends were different, and he seemed to have little time for girls, so our interactions in high school were primarily limited to football. James was neither popular or unpopular, just another student in the library studying his subjects and minding his own affairs or studies. Since there were no girls taking the building trades, he mainly only interacted with the other guys.

I felt prompted to write all this to show my lack of understanding about what makes a Marine or why he chose the more physical side of the Armed Forces. Perhaps it was because the recruiter was more convincing than the others, or the Marines guaranteed him the job as a radio operator when the other branches would not? Did the uniform impress James more than the nice white Navy uniform, or perhaps the Air Force blue? All of us young men had to register with the draft board which meant the distinct possibility of being drafted into the Army if one were physically qualified. Consequently, most of us volunteered to have some say in our futures and also in hopes of obtaining a better assignment. Perhaps the enticement for him was going overseas and being part of a detachment at an international embassy? Maybe it was as simple as the fact that communications, radio and radar and other methods, were of top interest to him. I certainly imagine that the Marine Basic (and perhaps even their job training) was more serious and demanding than that of the Air Force. Sometimes we in the USAF were looked down upon as the lesser service branch - thought of more like Boy Scouts.

James attended Radio Operator School (MOS-0621) for 16 weeks before reporting to his home or permanent base if Marines have such. Most are stationed on board ships as protection and to keep order. The Marine Radio Operator's number one responsibility was to contact and connect with distant stations and process and log messages. They received training in communication security and could make changes to frequencies or cryptographic codes. As a field radio operator advances in rank and experience, the next progression is training for Staff Sergeant through Corporal.

Although a versatile Marine like a radio operator can be utilized in any unit in the Corps, the communications man will not be used as infantry men or tankers, but if assigned to one of their units he will be fighting alongside of them carrying a portable radio and/or operating the vehicle-mounted one.

As I wrote these few lines attempting to discover why men choose different branches of the military, I realized that it was as pointless as trying to figure out why men choose the women they do for wives. From his daughter's description of James, as a father he was a genuinely nice, quiet, dedicated, hard worker and a loving person. Just like the majority, James had some adjustments to make after being and doing the USMC "stuff" every day for 6 years, but according to her he did not allow the Marine Corp's indoctrinations to carry over into his family life. She describes him as almost perfect: a man who never lost his temper or even showed the emotion of anger (as in raising his voice, etc.) whether he was at home or out in the field or cow pasture. In the home, his daughter said he was a good provider and a loving, respectful husband and father.

I can attest to the continuity of his character in latter life as well. After we were both honorably discharged from the services, we met and it was very apparent from our conversations that he was the same person and had never allowed his military service to influence him the way so many prior-service people do.

> It is tough for many prior service members to override the training and mindset that were previously critical to perform the tasks required and to our survival in the military! After being released from active duty, many are not able to readjust back to the "normal" life they had before enlisting. (The military is finally recognizing that addressing this area is vital to a healthy transition back to civilian life.)

Master Sergeant Charles Middleton: This Air National Guardsman was a great athlete who played quarterback on his high school football team. Charles graduated from a small high school in Middle Tennessee in 1953. To those of us who knew "Lefty" as we called him, he was an exceptionally good friend, a hard worker on and off the field, and a good leader with no negatives about him either in the classroom or at school. Although they did not win many games the year he played quarterback, it was not due to any shortcomings on his part. He took a great deal of abuse by his coach whom he was related to.

While collecting information on fellow service members, Lefty is one of the most honored members and a highly decorated one. He was stationed at McGee-Tyson Field in Knoxville, TN. His Tennessee Air National Guard Unit was called to active duty during the Berlin Crisis in 1961. As a showdown between the United States and Soviet Union, apparently John F. Kennedy decided to activate more than 65,000 Guardsmen, including the 151st Fighter Squadron at Knoxville, TN. Flying the F-104 jet aircraft, the Volunteer Guardsman participated in one of the largest deployments of jet aircraft in the nation's history in which 216 jets flew across the Atlantic Ocean within days after the mobilization.

> It is possible that the Bowe brothers took part in this operation. Airman First Class J.C. Bowe was a Flight Engineer on the KC-97 that did some refueling missions on these aircraft as they crossed the Atlantic, and Airman Second Class Bud Bowe was Loadmaster on the C-124 that flew supplies to the troops in Europe and other supporting locations for this mobilization and the Lebanon Crisis. He may have flown some of the missions when the Soviets constructed the Berlin Wall in August 1961. Guardsmen from many states, from both the Army and Air units, showed American resolve in the ensuing showdown by supplying food to Berlin.

Bobby Thompson was also a Marine and a fine football player and student at Bixby Seminary.

Corporal Claude Tindall is another veteran from this small school in Tennessee who volunteered his service to the U.S. Army and enlisted in the year 1954.

Alfred Thomas Bradford left our beautiful community in Tennessee and took the usual route for U.S. Air Force recruits to Houston Barracks in Nashville for testing. Al, as he is known by many friends in and around the town of Bixby, stated that listing the high school certificate on his entrance exam carried a great deal of weight with the Air Force at the time. It gave him many choices concerning his career, especially for someone in the state of Tennessee, and in the year 1949 he was selected to be a radio operator from the many positions he qualified for. After testing and being sworn into the Air Force, he was shipped by bus to AFTSK AFB, Mississippi.

Alfred told us that his basic training unit was part of President Truman's integration policy and his was one of the first integrated. He said there were three Southern Blacks that came into his Flight who were well behaved and finished with the rest. After graduation from Basic Training, Al was shipped to

Germany to attend The Air Force Radio Communication School in Munich, Germany, where he learned Morse Code. (Dot-dash-dot-dot, etc.) No doubt he was promoted to A/2C after school graduation, and then to A/1C as soon as he was eligible in the E-4 permanent pay grade on April 1, 1951. (The Airman First Class rank was a sergeant that year.) Alfred had enlisted for 2 years but was extended a year because of the Korean war. The war started while he was on furlough on the East Coast.

While he expected orders back to his old unit or Korea, Al was grateful that neither happened. He was in Camp Kilmer, New Jersey when discharged the 24th of September in 1952, after serving 3 years, 3 months and 10 days. His overseas time was 2 years, 1 month and 23 days. Bradford was awarded the following medals: Occupational, Good Conduct, Cold War, and National Defense. Alfred had also been a member of the TN National Guard in the Cookeville Unit prior to joining the Air Force. He had enlisted with them on March 1, 1948, in the HQ Co. 765th Tank Battalion and was released May 28, 1949. The Air Force recruiter had a difficult time obtaining Al's release and commitment to them.

Staff Sergeant Jessie J. "Clyde" Nichols, U.S. Air Force: Clyde was another childhood friend and a Cedar Hill School classmate of the Bowe brothers who was also a National Guardsman that served with Sgt. J. R. Thompson and I in the Cookeville Detachment. He traveled to Ft. McClellan for training but was perhaps in infantry or supply because I cannot recall him training in the tanks with our crew. On the Aggressor Force with me in July 1953 as well, Clyde was a good soldier and friend. After that, he served honorably in the Air Force for 20 years and raised a nice family out in Texas. He is now very much missed by his surviving family and in the Putnam County community.

Staff Sergeant First Class Pascal Jake Pulman: Jake, as we in the Bowe family all knew him, was not the military type of person and was against having his story in the book, but his surviving spouse (my sister Shirley) would like all to know what an exceptional husband, father, and grandfather he was. Born and raised in Chattanooga, TN, his father was a WWII veteran whom he very much admired. (His father brought back a souvenir from Germany and Jake spoke of it and his father often.) A couple of weeks into basic training, Jake developed pneumonia and was hospitalized for 14 days. This rocky beginning to his military time delayed his graduation almost a month.

His MOS (Military Occupational Specialty) was never exactly determined because there was no specific slot available that he qualified for. Given the delayed graduation from basic, they eventually just put him in an empty mail clerk position. Jake knew the job would require some typing and letter composing, among having to read and respond to questions about the job, and many other tasks he wasn't prepared for. Possessing the spirit of the thousands of Tennessean volunteers before him while waiting for a replacement, he used common sense to get the job done and came up with alternate ways to accomplish a job typically held by people who had at least an Associate Degree in Business or English Literature.

The top NCO's and the Commander took note of now Specialist E-4 Jake Pulman's work and accomplishments and recommended him for promotion while working out of his MOS. So, in short order he became Sargent E-5 in his 5th year while he was attending school at Tech and doing the mail clerk job. Even though he was being cross-trained into another MOS, he took the necessary test to be promoted to SFC (Sergeant First Class), Grade E-6, and went for all the recruiting persuasions to reenlist and make Master Sergeant (MSgt). While serving in the special unit, Jake attended college and received a bachelor's degree in business.

Sometime during his mail clerk job he met my sister, Shirley, who had some stories about the many sacrifices she made and things she had to endure while he attended college to obtain the knowledge to perform his civilian job after his release from active duty with the Tennessee National Guard. Sgt. Pulman received a special letter of commendation for being the top graduate out of his class of 200 in Basic Training. He attended and completed the Senior Division Reserve Officers Training Corps at Tennessee Technological University in Cookeville, TN. No one I ever knew, except those who played the political game, received a promotion to Sergeant First Class for sitting on his butt besides Jake.

PFC Ralph Short was another Tennessee volunteer who did his duty serving his country that did not want to be classified as a hero. He went to basic training at Ft. Jackson, then on to Quartermaster School in Virginia, and on to the Air Force base in Alaska for the duration of his enlistment where he guarded the ammunition dumps, including the missiles and nuclear warheads. This was an important base because of its proximity to Russia.

History of the Cookeville National Guard Unit

Years before WWII, the armories were constructed as places needed for training soldiers to defend their country, state, and communities. These became the National Guard Units as we know them today. I feel it important to also recognize the many Company Commanders of the Tennessee National Guard. These men are highly decorated and honorable heroes that served our country faithfully.

The information below about the Cookeville Unit of the National Guard was copied and taken from a publication entitled *The Fourth Annual Reunion Tennessee National Guard - Cookeville and Upper Cumberland Area*: The history of the Tennessee National Guard dates to the Battle of Kings Mountain when a bunch of rag-tag mountaineers from Tennessee crossed over the mountain and gave the British a whipping during the Revolutionary War. Of course, they were only a local band of State Militia back then.

- TN volunteers also fought the Indians in 1774 and fought in the Battle of New Orleans in 1813.
- Tennesseans fought in the 1818 War against the Seminole Uprising and again in 1836 during the Second Seminole War.
- 1887 marked the first time the Tennessee Militia was officially known as the TN National Guard.
- From the years 1923 – 1940, the Cavalry in TN was known as the 109th Cavalry Regiment; it was called up into federal service by Feb. 1941.
- The Cookeville units, and all of the others throughout the Upper Cumberland area prior, were mobilized and served during WWII.
- After the war, the members of these units did not return to the old armories when released from active duty because there weren't enough of them to do so. The armories remained empty until the formation of the new national guard units.
- The 765th Tank Battalion was the first formed in this area and their Headquarters and Headquarters Company became organized in Cookeville on February 6, 1947.

This book, <u>They Served to Keep the Peace,</u> only covers and records some events dating back to 1900 and up to and including 1960; any event after that will be in my forthcoming book of Navy stories.

Acknowledgments and Resources

Acknowledgments

- Chief Editorial Officer - Dana Fordice
- Others in the editing process: David Steen, Christopher Davis, Jamie Hartman, and others as well as myself - D. Ray Bowe.
- Cover Page: Jessie Nelson at Diversified Graphics
- Computer Technician: Tim Swiss - a companion and adviser who assisted in many ways
- I commend my brothers, Jesse C. and Buddy H., for their detailed input into their stories. We brothers shared the stories of good times and the worst.
- Thank you to my sisters Carol, Peggy and Shirley for their support and input.
- I deeply appreciate the encouragement from my nephew, Harold Ralph Maynard.
- David Steen for his contributions to the England and Bomb Competition stories.
- Thank you to Matthew Kris, my hometown librarian, for help with picture sizing.
- The leaders and sponsors at the Senior Citizens Center of our hometown.

Sources of Photos and General Information

- Some aircraft photos from the wall poster American Aviation 1947–2010 from Hobby Lobby.
- Most military aircraft photos and data came from the Encyclopedia of America Aircraft, forward by General Ira C. Eaker (Ret)
- All References to General Curtis Lemay were taken from the book Beyond the Wild Blue By Walter J. Boyne.
- The photo of the burning image of the C-119 came from the AP Charleston News.
- Photos of the 4-H livestock come from the Smith Co. Fair Newspaper, 2020.
- The barns and farmland photos were taken from the Farm Bureau mailer. Similar photos were used out of my subscriptions to Tennessee Magazine and the Farm and Ranch Magazines.
- The photos of the mules pulling implements and the wagon are of and used by permission from my neighbor.
- Some photos of military vehicles are taken from the American Legion Monthly Report. Other military vehicle images, including tanks, came from online sources such as Bing Images.
- The photos of me and my friends in Guam and on Mt. Home Air Force Base are authentic.
- Firsthand America: A History of the United States by Bernard/Burner/Fox-Genovese for information about the United States economy during the time when I was searching for a job after separation from the Air Force.

www.ingramcontent.com/pod-product-compliance
Lightning Source LLC
LaVergne TN
LVHW060037210926
848481LV00013B/3888